D0291118

WHAT LIGHT THROUGH YONDER WINDOW BREAKS?
More Experiments in Atmospheric Physics

CRAIG F. BOHREN
Pennsylvania State University

Foreword by
DAVID JONES ("Daedalus")

DOVER PUBLICATIONS, INC.
Mineola, New York

Copyright

Copyright © 1991 by Craig F. Bohren
All rights reserved.

Bibliographical Note

This Dover edition, first published in 2006, is an unabridged republication of
the work originally published in 1991 by John Wiley & Sons, Inc., New York.

International Standard Book Number: 0-486-45336-7

Manufactured in the United States of America
Dover Publications, Inc., 31 East 2nd Street, Mineola, N.Y. 11501

*Dedicated with gratitude and affection
to all my teachers, especially those who had
the greatest influence on me:*

*Robert F. Clothier, John O. Kessler,
Donald R. Huffman, and
the late Herold J. Miller.*

Contents

Foreword

*K*nowledge is power: or at least that's the modern idea. Science is valuable because it leads to technological progress, economic growth, and a more gadget-packed lifestyle. This may well be true; but to scientists like Craig Bohren it's rather beside the point. For science is the systematic exercise of curiosity. Learn a bit of science, and you'll gain a richer mental life. The world will make more sense and become a more appealing place in which to live.

For those whose main mental light breaks from the vacuous window of the TV screen, this may seem a revolutionary message. But let Craig Bohren get you into the habit of looking at the real world, and you'll find that even small things will repay notice and study. Anyone can learn the art of seeing and wondering: about dew on the inside of a windowpane, the momentary continued glow of a light bulb after the switch clicks off, oncoming vehicles miraged in a hot highway, or bubbles forming in a pan of water before it boils.

Simply noticing such things is mind-expanding on its own. Seeing that there is a question to be asked is even better, and trying to dream up an answer is better still. Craig Bohren's approach is best of all. Not only does he propose an answer, he sets up simple experiments to see if his answer is true. In his company you will measure the cooling rate of flasks of hot water wrapped in various materials. You will dust talcum powder on shaving mirrors, drop bus tokens into pans of diluted milk, and climb hills to measure the boiling point of water at the summit: all to test the truth of simple scientific ideas. Along the way you will enjoy the demolition of various common beliefs, such as that moist air conducts heat better than dry air, that the earth is nearer the sun in midsummer, and that the atmosphere traps the earth's heat like a blanket. Above all, you will learn something of the scientist's approach to the world, that powerful combination of practical curiosity and developed skepticism. Throughout this book, knowledge is pleasure!

David Jones
Physical Chemistry Department
University of Newcastle upon Tyne
Author of the "Daedalus" column in Nature
and "The Further Inventions of Daedalus"

Preface

*T*he modest success—as measured by the standards of popular science, not those of romance novels—of my book *Clouds in a Glass of Beer* encouraged me to write and my editor to publish this sequel. Most of the prefatory remarks in *Clouds* (as I call it for short) could just as well precede *Window*, but I shall not repeat them here.

Although I tried to make *Window* as self-contained as possible, to make it completely independent of its predecessor would have added too much bulk. Readers of this book who find my discussions of some topics—especially nucleation, humidity, and scattering—a bit terse are advised to consult *Clouds* for more details.

The message I have tried to convey in both books is that the physical phenomena of everyday life, especially those occurring in the atmosphere, are a source of endless delights to the eye and challenges to the mind, yet readily understandable to those who would make careful observations, do simple experiments, and reflect on them. Nothing I discuss requires specialized knowledge or expensive equipment.

I shrink from saying, as science writers often do, that this book is written for the layman, given that its antonym is priest; I dispense demonstrable truths, not revealed ones. Let me merely say instead that I write, although not exclusively, for those who do not make their living from science.

In the preface to *Clouds* I enjoined readers to be skeptical about statements made by authorities (including me), no matter how eminent. To my surprise, this raised the hackles of at least one reviewer. Yet I am not repentant. Indeed, I believe that as science becomes more complex and the pronouncements of scientific authorities become more obscure, even more skepticism is needed. A cursory glance at the history of science provides examples in abundance of confident and dogmatic, but wildly—even ludicrously—incorrect, assertions made by the scientific nabobs of the day. Fortunately, they were not armed with computers, so there were limits to the mischief they could cause. Today's nabobs are not so constrained.

Although I have tried to make only correct statements, one should never confuse effort with accomplishment.

Like its predecessor, this book is based on articles published in *Weatherwise*. Most of these articles appeared after the publication of *Clouds* in 1987. Heldref Publications generously granted me permission to use my articles. I have completely rewritten them, redone almost all of the figures, added material, and tried to weld the result into a coherent whole. Additions and emendations are the result of my own further thoughts—I never am completely satisfied with an explanation—as well as the comments and suggestions of readers, especially three whom I must thank by name.

Ever since I began writing the *Weatherwise* articles, initially with Gail Brown, then by myself, Duncan Blanchard at SUNY–Albany has been looking over my shoulder, giving valuable advice and making constructive criticisms. An alert reader, Charles Pierce, unearthed an error in my relative humidity calculations, which I have corrected. Among my readers who are not scientists by profession, Doug Beadle, a police dispatcher in Iowa, must be singled out as the reader who has caused me to give the most second thoughts to my explanations.

As a member of the largest meteorology department in the United States, perhaps even the world, I need not wander far from my office to find a knowledgeable colleague who can set me straight about some problem that has got my knickers in a knot. The colleague to whom I am most grateful is Alistair Fraser. So valuable are his criticisms that I wouldn't think of publishing anything without them. Indeed, even though intoxicated by gratitude, I would not be exaggerating by saying that I serve partly as his amanuensis and that the distinction between what is mine and his often is blurred. Much of this book has arisen out of our endless discussions, which are carried on by telephone well into the night and on weekends (my apologies to Dorothy Fraser).

Two colleagues contributed directly to this book by commenting on the drafts of chapters: Dennis Lamb (Chapter 1) and John Olivero (Chapter 10). My thanks to both of them, and my sympathies to John for having his office next to that of the most volatile member of our department. John also lent me a football that lay on my lawn for months, but I never got the frost-accumulation photographs I sought because of rain or snow or because I overslept on mornings when the grass was thick with frost.

Cliff Dungey critically read articles on which several chapters are based and also contributed to this book in ways mentioned in its text.

Two colleagues at Penn State beyond my departmental cloister who have contributed are Akhlesh Lakhtakia (engineering science and mechanics) and Herschel Leibowitz (psychology). Akhlesh helped me in my search for apt quotations for the heads of chapters, and Hersch patiently tutored me in visual perception. He also read the first draft of Chapter 15, which has been improved as a result of his comments.

Beyond Penn State I have colleagues who try to keep me on the straight and narrow, most notably Bill Doyle at Dartmouth, to whom I am deeply grateful for all that he has taught me. Bill's contributions are so many that if this book were a novel, he would be one of its main characters.

About one-fifth of *Window* was written while I was on sabbatical leave in the Department of Physics and Astronomy at Dartmouth during the academic year 1986–87. I thank Pennsylvania State University for granting me leave and John Kidder, Bill Doyle, and John Walsh for making it possible for me to enjoy it at Dartmouth.

A few chapters (or rather the articles on which they are based) were written during summers spent in the Life Sciences Division at Los Alamos National Laboratory, made possible by the generosity of Gary Salzman. I thank him and also my colleagues there, Roger Johnston and Shermila Singham (now at Sherwin-Williams).

Among the reviewers of *Clouds*, Jay and Naomi Pasachoff stand out as the ones who have caused me to redouble my efforts to purge my writings of all "infelicities." At *Weatherwise*, Pat Hughes and Jeff Rosenfeld have edited my articles with gentle but deft hands.

To Jearl Walker I am grateful for responding to my cries for help.

Gail Brown helped with some of the earliest articles, which finally are being incorporated into a book. Her photographs grace two chapters.

And speaking of photographs, I thank Lew and Jon Sheckler for advice on photography and for transforming my negatives into high-quality prints.

I say *tack så mycket* to two Swedish colleagues: to Carl-Gustaf Ribbing of Uppsala University for causing me to tighten some of my arguments in Chapter 7 and to Claes-Göran Granqvist of Chalmers University for his unwitting contributions to Chapter 9.

Although this book does not report research, it was written while what I call my research was supported by the Atmospheric Sciences Division of the National Science Foundation and by the National Aeronautics and Space Administration. I am as incapable of drawing sharp distinctions between research, teaching, and popularization as I am of respecting the boundaries between disciplines. Recently, I was delighted to learn that a result I obtained solely to support an argument made in a popular article has caught the attention of a French theoretical physicist, who has immortalized my result as "the Bohren theorem" despite its humble origins. I am also proud to report that Carl-Gustaf Ribbing cited the *Weatherwise* article on which Chapter 9 is based in a serious scientific paper.

Many thanks go to David Sobel at Wiley for his patience with my dilatoriness, for his advice, and for his encouragement. He also likes dogs.

Manoja Dayawansa took the data for the cooling curves in Chapter 12 and also labored mightily during an Easter vacation on one of the figures for Chapter 15.

As always, my deepest gratitude goes to Nanette Malott Bohren, or "poor Nanette" as she is known to our friends. Critic, copy editor, proofreader, photographer's assistant, and model for racy photographs, she dreams that each book will be my last.

Tŷ'n y Coed
Oak Hall, Pennsylvania
April 1990

What is the ultimate nature of matter? The question we know by now is meaningless. It would make laymen as well as physicists feel better to answer it—even as the idea of God makes some people feel better. How does the outside world work in a given context, approximately? That seems to be the sum and quest of human knowledge. It will give us as much power over the environment as we are competent to handle.
—STUART CHASE

All perception of truth is the perception of an analogy; we reason from our hands to our head.
—HENRY DAVID THOREAU

I once knew an old manufacturer who said: "All information is false." And he was right, for almost everything is exaggerated, distorted, or suppressed.
—ANDRÉ MAUROIS

Anyone who conducts an argument by appealing to authority is not using his intelligence, he is just using his memory.
—LEONARDO DA VINCI

There is little hope that he who does not begin at the beginning of knowledge will ever arrive at its end.
—HERMANN VON HELMHOLTZ

✳1

Window Watching

I must go seek some dew drops here
WILLIAM SHAKESPEARE: *A Midsummer Night's Dream*

Grading examinations can be a painful reminder of the yawning gap between what students appear to know—a call for questions is almost always met with silence—and what they actually know. It's usually tedious work, which I begin with a heavy heart and end with a headache.

At Penn State we give our graduate students a qualifying examination. The students hate to take it; the professors hate no less to grade it. We submit questions, they are sifted by a committee, the residue is inflicted on the students, then each question is graded independently by two professors. Once, I had to grade a question submitted by Toby Carlson, one of my colleagues. It lay on my desk for more than a week, and only after I had exhausted my excuses for not looking at this question did I reluctantly sit down to read it and the answers to it. The question was as follows: During the winter, dew often forms on windows, but always on their *inside* surfaces. Why?

I perked up when I read this. Suddenly, I was interested. Although I had seen dew on the insides of windows many times, I am ashamed to admit that I had never really thought about this before. Now I could think of nothing else.

More Is Not Always the Answer

As could have been predicted, the answers to Toby's question contained vague statements about "more humidity" inside houses than outside, although the students were not always clear if they meant greater *relative* or *absolute* humidity. *Absolute humidity* is the actual density (molecules per unit volume) of the water

1

vapor component of air. *Relative humidity* is the water vapor density relative to what it would be if it were in equilibrium with liquid water.

Relative humidities inside houses during winter, especially in the colder parts of the United States, are not high, as evidenced by sales of humidifiers, the sole function of which is to *increase* relative humidity. I learned this the hard way when some of our furniture cracked during one of our first winters in Pennsylvania. Then we bought a humidifier.

Firmer evidence that relative humidities inside houses in winter are usually less than those outside is easy enough to obtain. One morning when there was dew on the inside of our windows and the temperature inside our house was low, I measured relative humidities. The house hadn't been heated since the evening before, and no one had cooked or bathed since then. The dry-bulb temperature inside was 12°C (54°F); the wet-bulb temperature was five degrees lower, which corresponds to a relative humidity of about 48 percent. Outside, the dry-bulb temperature was 1°C (34°F) and the wet-bulb temperature was two degrees lower, which corresponds to a relative humidity of about 63 percent. Thus higher relative humidities inside cannot be necessary for dew formation on the insides of window panes.

Although the relative humidity I measured inside was less than that outside, the absolute humidity was greater. Occupied houses have various sources of water vapor: people breathing, sweating, bathing, and cooking. Outside air that drifts inside is heated and has water vapor added to it. Yet whether it is the relative or absolute humidity that is greater is irrelevant to Toby's question.

Why Does Dew Form on the Insides of Windows?

During the winter, temperatures inside houses are usually higher than those outside. Because of the continual transfer of energy from the warm interior of a house to its colder surroundings, the steady-state temperature profile in and near a window will be like that shown in Figure 1.1. The temperature of the inside surface of the window is *less* than the inside air temperature, whereas the temperature of the outside surface is *greater* than the

Figure 1.1
If the temperature inside a house is greater than that outside, and neither temperature changes with time, the temperature profile in and near a window will be as shown here. The rapid drop in temperature near both surfaces results from films of stagnant insulating air adjacent to them. Note that the inside window temperature is less than the inside air temperature, whereas the outside window temperature is greater than the outside air temperature.

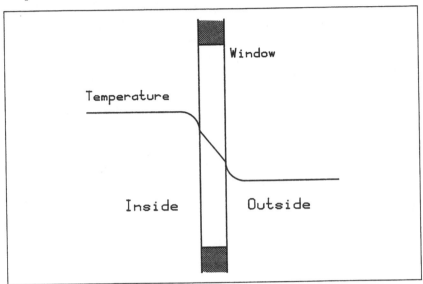

outside air temperature. These temperature differences and the concept of dew point are the keys to understanding why dew forms on the inside of windows.

The dew point is the temperature to which air must be cooled, at constant pressure, for saturation to occur. Stated another way, it is the temperature at which the rates of condensation and evaporation exactly balance; if air is cooled below the dew point, the balance is tipped in favor of condensation.

Usually, the air temperature is greater than, or at most equal to, the dew point. If the temperature of the outside window surface is greater than the outside air temperature, this surface is always above the dew point of the outside air; hence, dew cannot form on this surface. But the temperature of the inside surface may be lower than the dew point of the inside air; hence, dew may form on the inside surface.

The higher the relative humidity, the smaller the difference between the air temperature and the dew point. Although a higher relative humidity inside favors dew formation on the inside window surface, the essential reasons why dew forms on the inside rather than the outside surface are that in winter temperature decreases steadily from inside to outside a house and the air temperature is greater than the dew point.

For the outside surface of a window to be always warmer than the surrounding air, the window must not emit more infrared radiation to its surroundings than it absorbs from them (see Chapter 7 for more on infrared emission and absorption). Although this is probably true for most windows, which usually are vertical, it is not true for all surfaces, as you will discover in Chapter 9.

The qualifier *steady state* when applied to a temperature profile means that the profile does not change with time. If air warmer than the outside surface were to suddenly blow across a window, dew might form on it (although I never have observed this).

The process of dew formation on the insides of windows but not on their outsides may reverse during the summer, especially on windows of an air-conditioned house in a hot, humid environment. I have not observed this because I do my best to spend summers in cool and dry places.

Dew Patterns on Windows

Once your attention has been drawn to dew on windows, you are likely to notice a wealth of details worth thinking about. For example, you might see a pattern like that shown in Figure 1.2. For months during the winter this symmetrical pattern persisted on the windows of a house in which we lived when I was on sabbatical leave at Dartmouth College in New Hampshire. Dew formed on only the bottom part of the window, and more toward the edges than at the center. Let us consider each of these observations in turn.

Air in contact with a window is not static. It moves under the influence of *buoyancy* (see Chapter 11). Cold air is denser than warm air if both are at the same pressure. Inside warm air that comes into contact with a colder window is cooled and therefore sinks. Let us imagine following a parcel of air as it descends along the inside surface of the window (see Figure 1.3). The parcel cools and the window is warmed. The rate of this warming

Figure 1.2
This symmetrical dew pattern on a window of a house sheltered from both wind and direct solar radiation was remarkably stable.

is proportional to the difference between the window temperature and the parcel temperature. As the parcel descends, its temperature becomes closer to that of the window's, so the rate of warming of the window by the parcel decreases as it descends. Outside, the air is colder than the window; hence, air that comes in contact with it is heated and rises. The rate of cooling of the window by a rising air parcel is proportional to the difference between the temperatures of window and parcel. This difference is greatest at the bottom of the window and decreases with height because the rising parcel is warmed by the window. Thus, because of circulation of both inside and outside air along the window, its temperature is least at the bottom and greatest at the top.

To verify that temperatures are indeed lower at the bottoms of window panes, I measured temperatures with a thermopile, which is a stack of thermocouples in series. A temperature difference between the junctions of a thermocouple gives rise to a voltage dependent on this difference. Windows emit infrared radiation, and the higher their temperature the more they emit. By means of this infrared radiation absorbed by the thermopile,

Figure 1.3
Buoyancy-driven flow near a window causes the window temperature to vary vertically, being higher at the top than at the bottom.

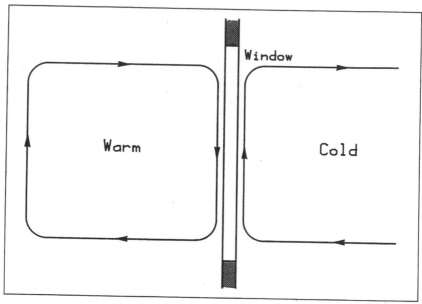

I could measure temperatures over the window, although relative rather than absolute ones. I did this at night so that sunlight wouldn't influence the results.

I first moved the thermopile horizontally, holding it close to the inside surface of the window. There was no noticeable change. But when I moved the thermopile vertically, the temperature it sensed changed markedly, the lowest value occurring at the bottom of the window.

Further evidence of lower temperatures near the bottoms of windows is provided by the size distribution of dew drops. I have noticed that often they are larger near the bottoms of windows, as in Figure 1.4.

What about the upward curve of the dew pattern toward the edges? This is more noticeable on small windows than on large ones. On my large picture window there was a noticeable upward curve of the dew line (the boundary between dew-covered and bare glass) near the edges but not so striking as that on the smaller adjacent windows. Again, I attribute this dew pattern to the imperceptible buoyancy-driven pattern of airflow over the window.

Figure 1.4
Drops are larger toward the bottom of this window, evidence that temperatures there are lower.

Air near the frames is retarded somewhat by them, just as water flows slowest at the edges of a stream and fastest in the middle.

I discussed my observations of dew patterns with Bill Doyle, a colleague at Dartmouth. This prompted him to examine his windows. He noticed patterns different from those on mine. Instead of curving upward near the window edges, the dew line curved downward. And near the edges there were thin strips of bare glass. He attributed this to heating of the window frames by solar radiation. Frames are not nearly so transparent to such radiation as is glass. Thus glass adjacent to the frames is warmer than it would otherwise be, sufficiently so that dew does not form there.

As I mentioned previously, the dew patterns I observed were remarkably stable. Every morning they appeared to be about the same, no doubt because our house was in a hollow, sheltered both from wind and from direct solar radiation (especially the front of the house where I made most of my observations). With the coming of higher temperatures in spring and a shift in the direction of the sun, the dew patterns changed markedly. I saw the same kinds of patterns on my windows that Bill saw on his.

The dew line in Figure 1.2 is therefore likely to be obtained only in a sheltered environment. Even the frame material may shape the dew pattern. It is not without consequence, I believe, that the frames shown in Figure 1.2 are aluminum. All else being equal, the surfaces of wooden frames will become warmer than those of aluminum frames because of the vastly different thermal conductivities of these two materials (for more on this, see Chapter 9), and this surely can affect the dew pattern. So although the pattern I observed is common and recurring, it is not universal. The definitive study of variations in dew patterns on window panes has yet to be done.

Frozen Dew and Frost

I began observing dew on my windows in the fall. At first the drops were liquid, but as temperatures fell I noticed that drops were sometimes frozen. Water drops that form on window panes below the freezing point do not necessarily freeze. A dew drop may exist for a while as subcooled water and then freeze because an ice nucleus either settles on it or is incorporated in it as the drop grows. At sufficiently low temperatures, water vapor condenses directly as ice, giving beautiful frost patterns such as that shown in Figure 1.5. Indeed, when I got up each morning during the winter, I could estimate the temperature from the frost pattern. Only when temperatures were below about $-12°C$ (10°F) did I see patterns like that in Figure 1.5. The obvious boundary is that between frozen droplets and frost. But what is the jagged line jutting well above the main body of the frost?

Wilson Bentley, a farmer from Jericho, Vermont, was the first and certainly the most famous photographer of snow crystals. In 1931, the American Meteorological Society published *Snow Crystals* (subsequently republished by Dover and still available) by Bentley and Humphreys, a collection of the best of Bentley's photomicrographs of snowflakes as well as a few of dew and frost. W. J. Humphreys, whose book *Physics of the Air* is a classic of atmospheric physics, wrote the text to accompany this collection. One of the photographs is of frost that formed in Bentley's initials scratched onto glass. It would seem that the jagged line shown in Figure 1.5 also formed in and around a scratch on the window. But how does the scratch favor the formation of frost?

8

Figure 1.5
When temperatures are low enough, water vapor will deposit on windows as ice. The obvious boundary is that between frozen water drops and frost.

This question also plagued Humphreys, and he took a few stabs at answering it. He recognized that the edge of a crack might cool more than the surrounding flat glass (for more on this see Chapter 9) but rejected this explanation as inadequate given that crystallization from liquids also occurs along scratches. One idea I have toyed with is that a scratch is a dustbin for condensation nuclei (tiny particles on which water vapor condenses) or ice nuclei (tiny particles that initiate the freezing of water), but this seems to be grasping at straws. My guess—unsupported by experiment—is that frost will form about as readily in a fresh scratch as in one old enough to have collected considerable microscopic rubbish.

The explanation that Humphreys appeared to favor most, as do I, invokes differences between evaporation from concave and convex surfaces. Suppose that a water droplet on a flat surface and a droplet nestled in an adjacent crack (see Figure 1.6) are

Figure 1.6
Water molecules escape less readily from a droplet in a crack (concave) than from one on a surface (convex). A water molecule at the surface of the concave droplet has more neighbors restraining it than a water molecule at the surface of the convex droplet. As a consequence, evaporation proceeds more slowly from the concave droplet, all else being equal.

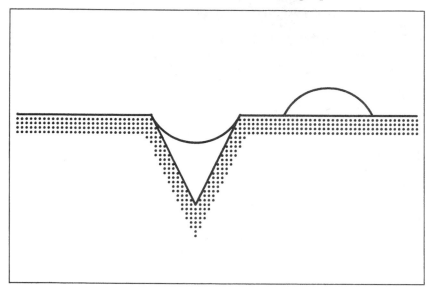

otherwise identical. From which droplet will evaporation be greatest? Evaporation increases with temperature but also depends on the radius of curvature of what is evaporating, especially for very small radii such as those of microscopic droplets. Because water molecules at the surface of a tiny convex droplet have fewer neighbors attracting them than do molecules at a flat surface, the rate of evaporation from the droplet is greater (all else being equal). The converse is also true: water molecules at the surface of a concave droplet have more neighbors restraining them than do molecules at a flat or convex surface. Hence the rate of evaporation from concave surfaces is less than that from convex ones. Dew (or frost) is water formed when condensation exceeds evaporation. Reducing evaporation by whatever means therefore can tip the balance in favor of condensation. So perhaps it should come as no surprise that both dew and frost (which can begin its existence as tiny droplets of frozen dew) may form more readily in scratches.

And speaking of scratching the surface, that is about all I have done as far as dew on windows is concerned. I have by no means exhausted all that can be learned from window watching. There is a lot left for you to puzzle over, such as why drops form in one place but not in another. Staring out of windows may be for some merely a waste of time, but for the observant it can be yet another window onto the fascinating world we inhabit.

*2

Interference Patterns on Garage Door Windows

What can we know, or what can we discern,
When error chokes the windows of the mind.
SIR JOHN DAVIES: *The Vanity of Human Learning*

*I*n Pennsylvania I walk to work. This gives me exercise and allows me to do some uninterrupted thinking. And I often see things that arouse my curiosity. During one academic year, however, I was on sabbatical leave at Dartmouth and had to drive to work. Although my body and soul suffered from want of stimulation on the twice-daily journey, I did extract some unexpected pleasure from it—as well as grist for this chapter.

Early in the fall, I would be home well before sunset. But as the days became shorter, I eventually had to drive home using headlights. One evening, as I was approaching my garage, I saw something that was both a delight and a puzzle. The windows of the garage door, illuminated by the headlights of my truck, reflected a series of brilliant colored bands. I can't recall ever having seen such a vivid example of *interference fringes*, similar to those displayed by oily puddles or soap bubbles. What puzzled me was why they could be seen in the garage door windows, which I knew were much too thick to give interference colors. The pleasure of seeing them was therefore diminished somewhat by my instinctive reaction that they couldn't exist.

My next thought was that the source of the fringes must be my windshield. However, I could also see them while standing alongside the truck, its headlights illuminating the garage door windows. These windows, the source of such striking patterns, were otherwise undistinguished. In the light of day, they were like countless other windows, but in the dark of night, they

13

transformed reflections of white headlights into a splash of colors. Why?

A Winter Night's Experiment

My observations of colors in garage door windows occurred during the winter term at Dartmouth, when I was teaching a freshman course on light and color. My students were required to write four essays, at least one of which had to be accompanied by photographs. One student, Andrew Mottaz, expressed an interest in interference colors. I told him about my garage windows and suggested that we do some experiments on them. So one bitterly cold night we set out to test the hypothesis that the colors seen in my garage windows resulted from interference in an imperceptibly thin film on the panes.

We set a camera, equipped with a telephoto lens, on a tripod at the end of my driveway. What we photographed first is shown in the top part of Figure 2.1. A hand, which blocks the glare of the headlights, is out of focus. Although you might think that this is merely an example of bad photography, it illustrates a point: if the window is in focus the fringes are not, and vice versa. Although they owe their existence to the window, they are not coincident with it, just as your image in a mirror lies behind it.

After taking several photographs of the window, we then washed half of it. On the cleaner half, the interference fringes vanished; this is shown in the bottom half of Figure 2.1. We had to scrub vigorously to get rid of the fringes, which attests to the robustness of the film.

Interference Fringes in Thin Films

Light is schizophrenic: it exhibits different personalities depending on the circumstances under which it is encountered. To understand interference phenomena requires coming to grips with light's wave personality.

A certain degree of *coherence* is necessary for interference. To *cohere* means to stick together, and the word is applied especially to entities that are alike. Two coherent beams of light stick together in the sense that their phases bear a fixed relation to each other. Consider two waves, equal in amplitude, propagating to-

Figure 2.1

A series of fringes is seen in this garage door window illuminated by the headlights of a truck (top). To show that these fringes owe their existence to something on the pane, the top half of the window was thoroughly cleaned, causing the fringes to vanish (bottom).

gether in the same direction. If their peaks coincide, they are said to be in phase; they interfere constructively and combine to give an intensity four times that of either one propagating alone. If the peak of one wave coincides with the trough of the other, they are said to be out of phase; they interfere destructively and combine to give zero intensity. Interference is the salient characteristic of waves, constructive and destructive merely being the two extreme degrees of interference; all intermediate degrees are also possible.

Now let us suppose that the two waves are fluctuating independently of each other, that is, they are incoherent. You can visualize this by imagining their sources to be turned on and off at random. At one instant the waves may happen to interfere constructively, whereas at another they may happen to interfere destructively. If the time over which this change occurs is small compared with the time it takes to observe the intensity of the combined waves, the combined intensity will be twice that of either individual wave. In this example light does not exhibit its wave personality: the two waves are observed not to interfere on average although they do so at each instant. If the waves fluctuate in step—that is, if they are coherent—they may give rise to observable interference.

Light sources of our ordinary experience are incoherent. If you are reading a book by the light of a single lamp and find it inadequate, you turn on another lamp. The resulting intensity of illumination on the printed page is just the sum of the separate intensities of each lamp. One lamp does not know what the other is doing; they are incoherent. But if coherence is necessary for interference, why is it that we can observe interference with ordinary light sources such as the headlights of my truck?

Although beams of light from *different* points of a light bulb are incoherent and hence do not interfere, light from the *same* point can be subdivided so as to give one or more beams that are coherent and thus can interfere. An example of this is shown in Figure 2.2, which depicts light reflected by a smooth slab. The incident light originates from a single point of a distant source. The reflected light may be considered to be a superposition of components: reflected at the first interface, transmitted after having undergone one reflection within the slab, transmitted after having undergone three such internal reflections, and so on. To

Figure 2.2
Light reflected by a slab can be decomposed into a set of component waves, only two of which are shown here. Because components 1 and 2 have propagated different distances before combining, they are almost out of phase.

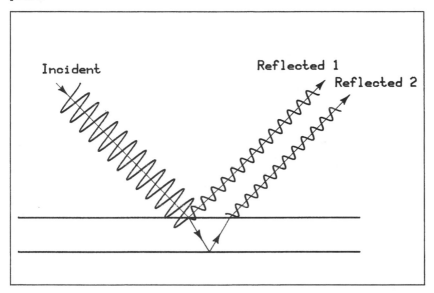

avoid clutter, only two components are shown. The observed reflected light is a coherent superposition of all the components. It is coherent because they are derived from the same source: if the incident light fluctuates, so do all the reflected components, and in the same way.

The intensity of the reflected light depends on the phase relations among its components. Although they all have a common source, they have different phases because they suffered different fates before combining. These phase differences depend on the thickness of the slab and its refractive index (i.e., its composition) as well as the direction of the incident light and its wavelength. In Figure 2.2 these quantities are such that the two components shown are almost out of phase and thus interfere destructively.

Because the phase differences among the various reflected components depend on wavelength, we can see interference colors if the source of illumination is white light. Interference among the reflected components may be constructive for a narrow range of wavelengths and destructive for all other wave-

Figure 2.3
Even though the light incident on an oil film on glass may be uniformly distributed over the visible spectrum, the light reflected is not so distributed because of interference. For the spectra shown here, the incident light is inclined five degrees from the perpendicular to the film. The solid line is for a film 0.35 μm (350 nm) thick, whereas the dashed line is for a film 2.0 μm (2000 nm) thick.

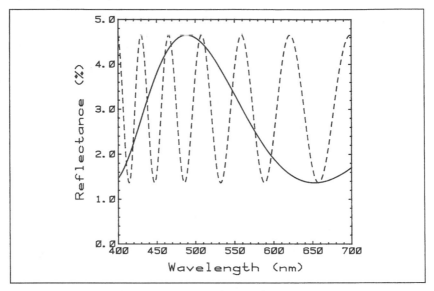

lengths within the visible spectrum. Although the incident light is white, that reflected is not: it has acquired color because incident light of some wavelengths—the ones for which the various reflected components interfere constructively—has been reflected more than light of others.

This is exemplified in Figure 2.3, which shows reflectance versus wavelength for an oil film on glass. When this film is 0.35 μm thick, the reflected light will be bluish-green even though the incident light is white.

Why Must Films Be Thin to Give Colors?

A discussion of interference in thin films is mandatory in every textbook on optics. Despite this, mention is rarely made of why films must be thin (i.e., comparable to the wavelengths of visible light) to give colors. The implication seems to be that this is so trivially obvious it hardly deserves comment. Perhaps it is obvious to some, but it wasn't to me, and since I may not be alone,

I included a calculation of reflection for a film 2 μm thick in Figure 2.3. Just as for the thinner film, the reflectance varies with wavelength, but there are several peaks in the reflection spectrum rather than only one. Interference certainly occurs in the thicker film, as indeed it should since my arguments did not hinge on any special film thickness. What is at issue is not whether there is interference in thick films but rather why we do not perceive colors in light reflected by them.

The key word in the previous sentence is *perceive*. Color is not objective. It is therefore senseless to divorce it from the human observer. The reflection spectrum for the film 2 μm thick displays a set of peaks, which could be observed with a suitable instrument. The human eye is not such an instrument. When we look at a source of light, we do not perceive the relative amounts of light in different wavelength intervals but rather a single visual sensation, which we have learned to describe by various names: red, blue, pink, white, and so on. We would perhaps call the light reflected by a film 0.35 μm thick bluish-green even though this light contains all visible wavelengths. So also does the light reflected by a film 2 μm thick, but no single wavelength or range of wavelengths predominates. Thus we perceive the light reflected by the thicker slab as white.

Why a Series of Regularly Spaced Fringes?

The photographs in Figure 2.1 show a series of more or less identical fringes. In the previous section I explained why interference in thin films can give colors. But my story is incomplete because I have not explained the regular spacing of the fringes seen in the garage window.

Each fringe in Figure 2.1 corresponds to light coming from a different direction (or small range of directions), hence a different direction of the incident light. Phase differences among the various components reflected by a film depend on, among other things, the direction of the incident light. One might therefore conclude that the series of fringes results from the range of directions of the incident light. This would indeed be possible if the light were nearly monochromatic, that is, confined to a narrow range of wavelengths. But to see as many fringes as are shown in Figure 2.1, even with monochromatic illumination,

19

Figure 2.4
The reflectance of very thick films for visible light of a given wavelength depends on the angle of incidence. The film thickness here is 150 μm (0.15 mm). The solid line corresponds to red light of wavelength 650 nm and the dashed line corresponds to green light of wavelength 550 nm.

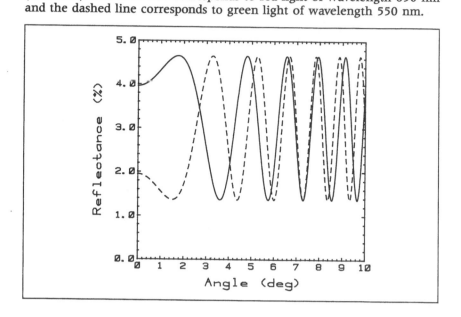

would require a film thickness vastly greater than the wavelength. This is shown in Figure 2.4. So I am faced with a dilemma. To explain colors, I must invoke thin films, but to explain a regularly spaced series of many fringes, I must invoke thick films.

Underlying everthing I have said up to this point is the tacit assumption that the films are uniform in thickness. A nonuniform thin film can certainly give rise to a series of colored fringes. For example, the colored patterns of soap films result from their variable thickness. A vertical soap film drains and is therefore thickest near its bottom. The film on my garage window was probably accumulated exhaust fumes. If this supposition is correct, there is no reason to expect the film to be uniformly thick, especially over an area as large as a window. Yet the regular spacing of the fringes seems to demand a film of regularly varying thickness (e.g., a wedge-shaped film). Unfortunately, it is difficult to imagine how a cloud of fumes billowing from an automobile idling in a driveway could be deposited as a wedge-shaped film on a nearby garage window. Nor can I envision any mechanism that would transform an irregular deposit into a regular one with time. So once again I have reached an impasse. There is one possible way out that I have not yet explored: the characteristics of the source of illumination.

Resolution of One Puzzle

Among the tools of the modern academic trade are thin, transparent plastic sheets onto which pearls of wisdom can be scribbled and subsequently projected onto a screen to be copied dutifully by students. My desk is littered with them. From time to time, while staring at them dreamily, I have noticed faint fringes. To the extent that I thought about them at all, I suppose I attributed them to interference in a thin film of air between sheets. But while musing over garage window fringes I remembered having seen fringes in a *single* sheet.

During a summer working at Los Alamos National Laboratory in the Life Sciences Division, I had ready access to the kinds of paraphernalia found in biological laboratories. I fished around for a microscope coverslip, a thin slab of glass about 0.15 mm thick, and studied it carefully in my office. It displayed a series of regularly spaced, faintly colored interference fringes.

This caused me a night of fitful sleep. I argued in a previous section that interference colors are not possible in films much thicker than the wavelengths of visible light. Yet I seemed to have evidence to the contrary before my very eyes. After much stewing, I finally remembered having seen a spectrum of the light from a fluorescent lamp. We call such light white because this is the sensation it evokes. But its visible spectrum is not uniform. It is punctuated by at least two sharp spikes, one corresponding to a green, the other to a violet—the colors I had seen in the coverslip when illuminated by fluorescent light. When I looked at the coverslip in sunlight or light from an incandescent lamp, the interference fringes vanished.

Interference fringes can be seen in thick films if the spectrum of the source of illumination is sufficiently narrow (see Figure 2.4). Although fluorescent light is far from monochromatic, it contains two intense monochromatic components jutting above a more or less uniform background, and it is these that give rise to the pastel interference fringes seen in films as thick as 0.15 mm.

There is a limit beyond which one cannot see interference colors in thick films illuminated by fluorescent light. For example, a microscope slide 1.23 mm thick does not exhibit them in such light, although it does in monochromatic light.

What little I could find out about the spectra of automobile headlights did not make me smile. I could find no spikes, which seemed to rule out peculiarities of headlights as the source of the many regularly spaced fringes seen in my garage windows.

Resolution of a Mystery

After I had exhausted my stock of possible explanations of the details of the fringes seen in my garage window, I cried for help. Although I did not doubt that the fringes were caused by a deposit on the panes—the experiment we did proved that—I did not understand to my satisfaction why there were so many fringes, why they were so vivid, and why they were so evenly spaced and symmetric.

Jearl Walker, who writes "The Amateur Scientist" for *Scientific American* and is the author of the celebrated *Flying Circus of Physics*, responded to my appeal. He had written about the fringes I observed in his August 1981 column. They are variously called Newton's diffusion rings, Whewell's fringes, and Quetelet's rings. Although they are indeed an interference phenomenon, they have nothing to do with thin films but rather are caused by particles on a thick film, either a glass slab or a mirror with its back surface silvered.

A particle on a window illuminated by a beam scatters light toward the back surface of the window, and part of this scattered light is reflected to the observer. But light from the incident beam also is reflected by the back surface and illuminates the particle, which scatters some of this reflected light to the observer. Interference between these two beams with different histories—scattered by the particle, then reflected by the glass; or reflected by the glass, then scattered by the particle—is the origin of the beautiful colored fringes I saw.

My friend at Dartmouth, Bill Doyle, went back to the house where we had been living and, after prudently warning the owners beforehand so as to avoid a whiff of buckshot, illuminated the garage door windows with a flashlight. This convinced him that I had indeed observed diffusion rings. Then he set up a demonstration himself using talcum powder sprinkled on a shaving mirror.

Our garage door windows in Pennsylvania are too high to be illuminated by headlights, so I used a flashlight. And sure enough, I saw vivid colored rings. Without much effort, you can see them in any dusty mirror or piece of glass. I am therefore puzzled by why I had never seen them before and embarrassed that I hadn't known about them. But I am pleased that I was not content to merely ascribe the rings I happened upon to interference and be done with them.

*3

Window Watching and Polarized Light

But, soft! What light through yonder window breaks?
WILLIAM SHAKESPEARE: *Romeo and Juliet*

A few years ago on a long flight to Hawaii I saw something that provides more grist for the mill I have been grinding in the preceding two chapters: what can be learned from the seemingly idle pastime of staring at windows. I was not staring directly at my airplane window but rather at its image reflected by my reading glasses, which I had taken off for a moment. They were dangling from my fingertips when suddenly I became aware of colors reflected in the lenses.

What I saw in the reflected airplane window was, like the fringes in garage door windows discussed in the previous chapter, a kind of interference phenomenon. I chose the words *a kind of* deliberately: to properly attribute colors in airplane windows to interference requires more care than is usually taken.

A friend once mentioned casually, much to his regret, that the colors seen when ice is interposed between crossed polarizing filters are interference colors. "They are not," I snapped. This occurred when I was at Dartmouth, and I told my friend there, Bill Doyle, about my encounter with a benighted soul who thought that colors in ice result from interference. Bill spouted and fumed as much as I had—perhaps even more. This confirmed the appropriateness of my earlier vehement response. Out of curiosity, however, I consulted various books. The proffered explanations of colors in transparent crystals between crossed polarizing filters were often nearly unintelligible. To the extent that I understood these explanations at all, their authors seemed to be saying that such colors result from interference between two

25

beams polarized at right angles to each other. I was thunderstruck by such heresy. To understand why, we must delve into the nature of polarized light.

Elliptically Polarized Light

The oscillating electric and magnetic fields of a beam of light—an electromagnetic wave—lie in a plane perpendicular to the beam's direction of propagation. It is easy to be confused about which direction is meant in discussions of polarized light because how it interacts with matter depends on two mutually perpendicular directions: the direction of propagation of the light and the direction of its associated field.

Let us imagine going to a particular point on a beam and following the electric field there as it oscillates in time. We may represent this field by a directed line segment, its length proportional to the strength of the field and its direction that of the field. Although this line segment lies in a plane, its direction and length are continually changing. Its tip therefore describes a curve. If this curve has a fixed shape, the light is said to be completely polarized. The most general fixed curve possible is an ellipse, called the *vibration ellipse* (Figure 3.1); hence the most general state of complete polarization is elliptical. Beams are in different states of polarization if their vibration ellipses differ in one or more particulars: orientation, ratio of minor to major axes, or even the sense in which the ellipses are traced out (clockwise or counterclockwise).

Both linear and circular polarization are particular kinds of elliptical polarization because straight lines and circles are particular ellipses. A line is an ellipse with a minor axis of length zero, and a circle is an ellipse with equal minor and major axes.

Light is unpolarized if the successive vibration ellipses traced out over many periods of oscillation of the field exhibit no regular pattern. This is difficult to represent graphically, but if we were to try to do so, the single ellipse in Figure 3.1 would be replaced by a childish scribble of many uncorrelated ellipses.

A beam of light may be treated as if it were the sum of two beams, one polarized and the other unpolarized. The ratio of the intensity of the polarized component to the total intensity is called the *degree of polarization*. Completely polarized (100 percent)

Figure 3.1
At any point along a beam of polarized light, the associated electric field oscillates in time so as to describe an ellipse in a plane perpendicular to the beam.

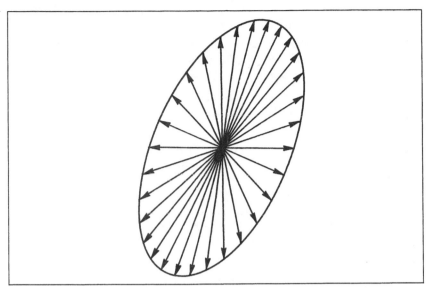

and completely unpolarized (0 percent) beams are limiting extremes. Careful measurements would probably always detect a departure from complete polarization in light we call polarized and a minute degree of polarization in light we call unpolarized. All light is therefore *partially* polarized to varying degrees.

Fresnel-Arago Laws

In 1819, two Frenchmen, Augustin Jean Fresnel, an engineer, and Dominique-François Jean Arago, an astronomer, published a landmark paper on polarized light. On the basis of experiments, they came to several conclusions, usually called the Fresnel-Arago laws. Only one of these laws, the first, concerns us here: "In the same condition in which two rays of ordinary light seem to destroy each other mutually, two rays polarized at right angles or in opposite senses exert on each other no appreciable action."

What this means is that two beams propagating together in the same direction, but linearly polarized at right angles to each other, cannot interfere. That is, the intensity of the two beams combined is always the sum of the intensities of each beam acting as if it were propagating alone.

Now you can better understand my shock at reading that colors in transparent crystals between crossed polarizing filters result from interference between beams polarized at right angles to each other. To have accepted this as literal truth would have required me to dislodge from my mind one of the moss-covered cornerstones of optics. Although I am always ready to reject as false what I have previously thought to be true, I am reluctant to do so on the authority of textbook scribblers. Yet their apparently heretical explanations can be brought into step with holy writ by taking a bit of care.

Transformations of Polarized Light

The only reason that the polarization state of light is worth contemplating is that two beams, otherwise identical, may interact differently with matter if their polarization states are different. Because of this, when light of one type of polarization interacts with matter, it generally is transformed into light of a different type.

To determine how the polarization state of a beam of light is transformed when it interacts with matter, it is convenient to resolve its electric field into two components perpendicular to each other (Figure 3.2). We may call these components vertical and horizontal, although, of course, these designations have no absolute meaning.

The vertical component oscillates through zero from a maximum pointing up to a maximum pointing down. Similarly, the horizontal component oscillates through zero from a maximum pointing right to a maximum pointing left. At any instant, the total field is the sum of these two fields.

The largest value of an oscillating field is called its *amplitude*. The ratio of the amplitude of the vertical component to that of the horizontal component determines the nature of the vibration ellipse. But this is not all—the *phase* difference between the two components is also a determinant.

Just as the two meanings of direction applied to polarized light are a source of possible confusion, so also are the two meanings of phase: the phase difference between two different beams with the same polarization and the phase difference between two different (perpendicular) field components of the same beam. It is the second meaning I have in mind here.

Figure 3.2
Any oscillating electric field lying in a plane can be resolved into two mutually perpendicular components, one vertical, the other horizontal.

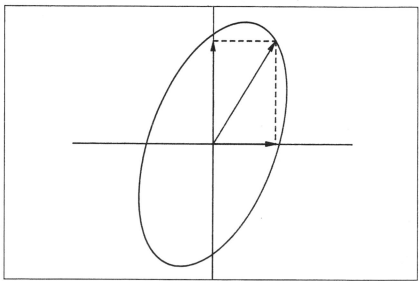

Two components are in phase if their greatest (positive) values occur at the same instant. (The vertical component is positive when pointing up; the horizontal component is positive when pointing right.) In this case, the field oscillates along a line the orientation of which is determined by the ratio of the two amplitudes (i.e., the light is linearly polarized). Two components are 180 degrees out of phase if the vertical component has its greatest positive value when the horizontal component has its greatest negative value. Again, the light is linearly polarized. These are just two extremes. All possible states of elliptical polarization correspond to all possible ratios of the amplitudes of the two perpendicular components and all possible phase differences between them.

To determine how the state of polarization of a beam changes when it interacts with matter, we resolve its electric field into components linearly polarized at right angles to each other. The amplitude and phase of each component are changed by interaction but, in general, by different amounts. After interaction, we recombine the two components to determine the new vibration ellipse.

Polarizing filters, such as those in sunglasses, are almost completely transparent to light linearly polarized in a particular direction, called the *transmission axis*, and almost completely opaque to light linearly polarized perpendicular to this direction. Suppose that a beam is incident on a polarizing filter. We resolve the field of this beam into two perpendicular components. The component parallel to the filter's transmission axis is transmitted without appreciable change in amplitude, whereas the amplitude of the component perpendicular to this axis is reduced to almost zero. Thus the light emerging from the filter is linearly polarized along the transmission axis.

Polarizing filters transform amplitudes, whereas *retarders* transform phases. Crystals and airplane windows are examples of retarders. A (linear) retarder has slightly different refractive indices for light of different states of linear polarization. Suppose that a beam is incident on a retarder. We resolve the field into two perpendicular components, one along the retarder's *fast axis* (smallest refractive index) and the other along the retarder's *slow axis* (largest refractive index). If these two components are in phase entering the retarder, they will not be in phase, in general, emerging from it. The phase difference introduced by the retarder is called its *retardance*.

Now we can determine what happens to light incident on two crossed (i.e., transmission axes are perpendicular) polarizing filters between which a retarder is interposed. The initial filter linearly polarizes the incident light. This light can now be resolved into components along the slow and fast axes of the retarder (see Figure 3.3). Although these two components are in phase entering the retarder, they are not necessarily in phase emerging from it. Thus the light incident on the final filter is elliptically rather than linearly polarized.

In general, the components along the fast and slow axes have unequal components along the transmission axis of the final filter (components perpendicular to this axis are not transmitted). Without the retarder between the filters, these two components would be equal in magnitude but oppositely directed; hence, their sum would be zero and no light would be transmitted by the final filter. With the retarder in place, however, some light is transmitted, the intensity of which depends on the retardance. In turn, the retardance depends on wavelength. So if white light

Figure 3.3
Light polarized vertically is incident on a retarder (top). The field of this light can be resolved into perpendicular components along the fast and slow axes of the retarder. These two components are initially in phase. Upon transmission by the retarder, each component suffers a different phase change. The result is that elliptically polarized light emerges from the retarder (bottom). This elliptically polarized light is then incident on a polarizing filter with horizontal transmission axis. Only the projections of each of the two components onto this axis are transmitted by this filter.

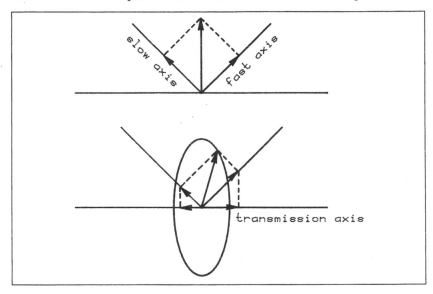

is incident on a polarizer-retarder-polarizer sandwich, the light emerging from it may be colored: light of some wavelengths is transmitted more than light of others.

To summarize the previous arguments in different words, we may say that the two perpendicular components emerging from the retarder themselves have components along the transmission axis that interfere. In this sense, and only in this sense, can beams perpendicularly polarized be said to interfere. To omit this qualifier is a misdemeanor, a violation of the Fresnel-Arago laws.

In scanning various books, I found only one author who recognized (implicitly) that ascribing colors to interference of perpendicularly polarized beams is slightly shady. As could have been predicted, this was Robert W. Wood, one of the giants of optics, whose *Physical Optics* is a classic. On page 346 of the third edition, I found the following, fittingly enough immediately below the Fresnel-Arago laws: "two beams of light, polarized at right angles, are capable of uniting into a circular or elliptical vibration, so that we can consider interference as taking place *in this sense"* (emphasis added). (This caveat was missing from the second edition, so either Wood or his customers balked at seeing the Fresnel-Arago laws, which prohibit beams polarized at right angles from interfering, followed by an apparent violation of the first of these laws.)

Although with effort one can reconcile the Fresnel-Arago laws with seemingly contradictory statements about interference of perpendicularly polarized beams, I prefer a concise alternative interpretation of the colors seen in a polarizer-retarder-polarizer combination, one that seems so simple and natural that I have never entertained another.

Light emerging from the initial filter is polarized perpendicular to the transmission axis of the (crossed) final filter; hence, no light would be transmitted by it if there were no interposed retarder. But the retarder transforms this linearly polarized light into elliptically polarized light, which is then incident on the second filter. At least some of this light is transmitted, the amount depending on the retardance, hence on the wavelength. Note that no mention is made of interference. By simply appealing to the *transformation* rather than to the *interference* of polarized light, we avoid running afoul of the Fresnel-Arago laws.

The interference interpretation has the virtue that the mathematical expression for the intensity of light transmitted by a polarizer-retarder-polarizer combination is similar to that for the light reflected by a thin film. Yet this unification comes at the cost of possible confusion, and not only because of the Fresnel-Arago laws. In the previous chapter I emphasized that interference colors are seen only in thin (comparable to the wavelength) films. Airplane windows certainly could not be described as thin.

The first question to ask when confronted with an explanation that invokes interference is, what is interfering with what? If this is not made clear, the purported explanation is no more than an incantation. When I discussed colors in thin films, I attributed them to interference between light reflected because of the top surface of the film and light reflected because of the bottom surface. A phase shift between these beams is introduced because they propagate different distances. Although I did not say so, these were *optical* distances, the product of a physical distance (i.e., the film thickness) and a refractive index (i.e., the refractive index of the film).

A phase shift is introduced between two perpendicularly polarized components of light transmitted by a retarder because the refractive index for each component is slightly different. This phase shift depends on the optical path difference. Here the optical path difference arises from a difference in refractive indices, whereas for the thin film it arose from a difference in physical paths. Even though a retarder may be physically thick (i.e., many wavelengths thick), the optical path difference for light polarized along the fast and slow axes may be comparable to that for the component beams reflected by thin films because of the small difference in refractive indices for light of these two polarizations. That is, a large physical distance can be compensated for by a small refractive index difference.

Although colors in thin films and in airplane windows are indeed brothers under the skin, establishing their precise genealogy requires considerable spadework.

Back to Airplane Windows

Unpolarized sunlight becomes partially polarized because of scattering by atmospheric molecules and particles, yet another example of how the polarization state of light is transformed when it interacts with matter. The skylight incident on my airplane window was therefore partially polarized, the atmosphere itself serving as the initial polarizing filter. The window was a retarder, the retardance of which varied from point to point. This retardance pattern was a consequence of the nonuniform distribution of stresses, hence strains (minute displacements), in the window, which was made of plastic, not glass. Thus all the ingredients were at hand for seeing colors. For the knowledgeable air traveler, polarizing filters can provide visual delights as well as inscrutability.

While looking at airplane windows through polarizing sunglasses or the polarizing filter on my camera, I have seen beautiful colored bands. Yet the windows were illuminated by skylight, which might cause you to puzzle over how multicolored bands can be obtained from a source of blue light.

We call skylight blue even though it isn't pure blue. That is, it is not light of a single wavelength or narrow range of wavelengths in the blue (itself an imprecise term) but rather light of all wavelengths. Indeed, your reaction upon seeing a measured spectrum of skylight for the first time would likely be, where's the blue? Blue is merely the perceptually dominant component of skylight, not the sole one. Even the best blue sky, that near the zenith, is composed of light of all wavelengths. When I observed my airplane window, I was not looking at the zenith sky. If I had been, my thoughts would have been on more urgent matters. Fortunately, my airplane was flying more or less level, so the backdrop for my window was much of the horizon sky, brighter than the zenith sky but not nearly so blue.

The light incident on the window need be neither white nor completely polarized to give rise to colors, although it cannot be monochromatic or unpolarized. To show this, I calculated the spectrum of light transmitted by a retarder illuminated by light with degree of polarization 75 percent. This is high, but not unrealistically so, especially at high altitudes. I made a guess at the spectrum of the incident light, not as blue as the bluest sky but not as white as sunlight either. The results are shown in Figure 3.4.

I tried various retardances to see what colors resulted, and indeed I could reproduce the colors I observed. In particular, I convinced myself that it was not possible to obtain a good red. The two colored bands I observed were yellowish-green and purple. The latter is not a spectral color (i.e., there is no wavelength corresponding to the sensation purple) but rather a mixture of violet and red light, just the kind of mixture I calculated. The violet component arises from incident illumination rich in violet light and the red component arises from a retardance favoring the transmission of such light.

Colors in airplane windows are best near the horizon or below it and fade above it. I attribute this fading to two factors: with decreasing zenith angle the illumination becomes bluer and the degree of polarization (assuming a sun high in the sky) decreases. The degree of polarization of skylight is greatest at 90 degrees from the sun.

Up to this point I have had in mind only colors in airplane windows seen with the aid of a polarizing filter such as polarizing sunglasses. Yet I began this chapter by recounting an observation of colors seen in the reflection of an airplane window in the

Figure 3.4
Light with the spectrum shown by the dotted curve, and partially linearly polarized with degree 75 percent, is incident on a retarder. The light transmitted by the retarder is observed through a polarizing filter. Depending on the retardance, the transmitted light can be either yellowish-green (solid curve) or purple (dashed curve).

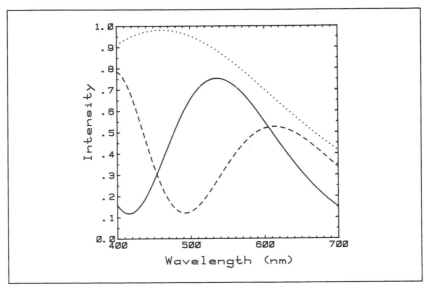

lenses of ordinary reading glasses. The second polarizing filter seems to have been absent. Or was it? I'll let you stew over this yourself until you read the next chapter.

*4

Fame from Window Watching: Malus and Polarized Light

In this light my spirit suddenly saw through all
JAKOB BOEHME

Window watching, a recurring theme in previous chapters, gave a measure of lasting fame to Etienne-Louis Malus (1775–1812), a French engineer. While looking through a calcite crystal at the setting sun reflected in windows of the Luxembourg Palace in Paris, he discovered a mechanism for polarizing unpolarized light, one that previously had escaped notice. This discovery, which bears on a puzzle I only partly unravelled in the previous chapter, was of sufficient moment to ensure for him a high rank among the investigators of polarized (a term he coined) light. Malus was not only an astute observer, he also did some simple experiments—and so can you.

Polarization upon Reflection

You can obtain a collimated beam with a slide projector and a black slide with a hole in it. Prop up the projector so that it obliquely illuminates a sheet of glass lying in front of it. The beam reflected by this sheet of glass is incident on another sheet that reflects the beam onto a screen (see Figure 4.1).

The angle between a beam and the direction normal to a surface it illuminates is called the *angle of incidence*; the plane defined by these two directions is called the *plane of incidence*. For this experiment, the angle of incidence onto the sheet should be about 57 degrees (i.e., the back of the projector should be tilted

Figure 4.1
A collimated beam from the projector is reflected by two glass sheets onto a screen (top). As the second sheet is rotated, the brightness of the beam spot on the screen decreases markedly (bottom). (White paper was taped to the back of the second glass sheet to make the beam more visible.)

about 33 degrees above the surface on which it rests), the exact value depending on the glass you use. Wedges will help you make fine adjustments.

The angle of incidence the first reflected beam makes with the second glass sheet also should be about 57 degrees. Keeping the second angle of incidence as close to this value as possible, rotate the sheet. You will discover that the brightness of the image projected onto the screen changes markedly. When the two planes of incidence are perpendicular, the image may even vanish (I say *may* because it is difficult to get all the angles just right). In the jargon of polarized light, the first glass sheet is called a *polarizer* and the second an *analyzer*; the first produces polarized light, the second detects it.

To understand this experiment, we have to consider what happens when unpolarized light is incident on the optically smooth interface between dissimilar media (air and glass in this experiment). No surface is smooth on an absolute scale: the qualifier *optically* signals that the appropriate measuring stick is the wavelength. Thus the bumps on an optically smooth surface are much smaller than the wavelength of the light illuminating it.

Light from the projector is unpolarized, which you can verify by looking at it through a polarizing filter (e.g., polarizing sunglasses) while rotating it. The electric field of the projector beam lies in a plane perpendicular to the direction of propagation. We can resolve this field into two components (see the previous chapter)—one parallel, the other perpendicular to the first plane of incidence. We may consider the beam incident on the first glass sheet to be composed of two equally intense beams linearly polarized at right angles to each other and propagating independently in the same direction. The fate of one beam upon reflection does not affect that of the other.

At near-normal incidence, the reflectances of the two beams are small (about 5 percent) and nearly identical. But as the angle of incidence is increased, the reflectances diverge, the difference between them being greatest at about 57 degrees, for which angle the beam polarized parallel to the plane of incidence is not reflected at all (see Figure 4.2). Thus at this *polarizing angle* (or Brewster angle), the reflected light is completely linearly polarized perpendicular to the plane of incidence. It is partially lin-

Figure 4.2
The reflectance of glass for incident light polarized parallel to the plane of incidence (solid curve) is different from that for incident light polarized perpendicular (dashed curve) to the plane. Note that the vertical scale is logarithmic.

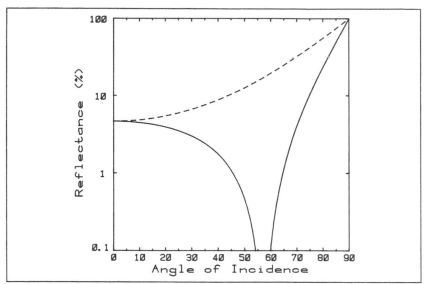

early polarized for other directions of incidence except normal and glancing.

Now we are better equipped to understand the experiment. Light was incident on the first glass sheet at the polarizing angle; hence, the reflected light was polarized perpendicular to the first plane of incidence. When we rotated the second glass sheet so that the second plane of incidence was perpendicular to the first, the light incident on this sheet was polarized parallel to the second plane. As we have seen, the reflectance of parallel polarized light vanishes when incident at the polarizing angle.

Before returning to Malus, I shall dwell a bit on the microscopic interpretation of our experiment. The following section is somewhat advanced, but you may pass over it without losing the essence of this chapter.

The Scattering Interpretation of Polarization upon Reflection

Regardless of the media involved, the polarizing angle has the peculiar property that it is the angle of incidence such that the reflected light makes an angle of 90 degrees with the transmitted (refracted) light. Can it merely be a coincidence that when unpolarized light illuminates single atoms or molecules, the light scattered by them at 90 degrees is, like light reflected at the polarizing angle, completely linearly polarized? This similarity gives rise to the scattering interpretation of polarization upon reflection, although it is sometimes presented with misgivings, as if it were merely suggestive rather than rigorously correct.

Once, after Bill Doyle (a leading character in previous chapters) had given the scattering interpretation of polarization to his physics class at Dartmouth, a student came to him and exclaimed, "Professor Doyle, the explanation you gave us today is wrong. It says so right here in this book." As Doyle told me, "This immediately got my attention." Piqued by his student's remark, he set out to show that there is nothing fundamentally wrong with the scattering interpretation. The fruits of his investigations eventually appeared in *American Journal of Physics* (Vol. 53, 1985, p. 463). Although I highly recommend his paper to anyone who wants to understand polarization upon reflection, understanding it demands more mathematical knowledge than does this chapter. I shall therefore try to give the essence of Doyle's arguments.

An illuminated piece of glass, say, is an array of a vast number of tiny dipolar antennas, each driven by the incident light and, as a consequence, radiating (i.e., scattering) waves in all directions. The superposition of all these waves nearly cancels except in two special directions, those given by the laws of reflection and refraction. To obtain the curves in Figure 4.2, I used the Fresnel equations, which specify how much light is reflected when light polarized parallel or perpendicular to the plane of incidence is directed at an arbitrary angle onto a smooth interface. Doyle showed that these equations can be factored into two terms: the scattering pattern of a single dipole and an array factor. The latter enters because the dipoles (atoms or molecules) form a phased array, meaning that there are definite phase relations among the waves radiated by all elements of the array.

The array factor is the same for both polarization states. Although this factor is never zero, the single dipole factor is zero for incident light polarized parallel to the plane of incidence when the angle between the reflected and transmitted light is 90 degrees. Scattering by a molecule (strictly, a spherically sym-

41

metric one) vanishes at a scattering angle of 90 degrees if the incident light is polarized parallel to the scattering plane, which is defined by the directions of incident and scattered light. For the problem of reflection at an interface, the plane of incidence is the macroscopic version of the (molecular) scattering plane. Thus the polarization of skylight and of light reflected by glass originate from the same underlying mechanism: the polarization state of light is changed when scattered by molecules, be they in air or in glass. The difference between them is that molecules are more densely packed in the latter. Whereas air is a random array of scatterers, glass is a coherent array. Although this leads to differences between them, many of their observable properties are determined solely by the behavior of their individual constituent molecules, which is similar for both.

Why Brewster Angle, Not Malus Angle?

Malus made his observations while looking through calcite at sunlight reflected by windows. The remarkable properties of this crystal (also called Iceland spar) were discovered by Erasmus Bartholinus (1625–92), a Danish physician. In a short memoir published in 1669, he wrote, "As my investigation of this crystal proceeded there showed itself a wonderful and extraordinary phenomenon: objects which are looked at through the crystal do not show, as in the case of other transparent bodies, a single refracted image, but they appear double." This is the first description of *double refraction*, which is more or less synonymous with *birefringence*. Although airplane windows are birefringent because of stresses, I would hesitate to call them doubly refracting because they do not give the kinds of double images observed with calcite.

The difference between airplane windows and calcite lies in their degree of birefringence. As the term implies, a birefringent material has two refractive indices, which are different for light of different (linear) polarization states. For calcite, the relative difference between its two refractive indices is about 10 percent, whereas for airplane windows I estimate it to be about 1/100th of 1 percent, not enough to give perceptible double images but enough to give other observable consequences, most notably colors.

Unpolarized light incident on calcite may be split into two well-separated beams polarized at right angles to each other.

Moreover, depending on its direction of incidence, linearly polarized light may or may not be transmitted by calcite. This capability of calcite to transmit light of different polarization states quite differently enabled Malus to make his famous observation, which he soon followed with experiments. On the basis of these he was able to write in 1809 that "the angle at which light experiences this modification when it is reflected at the surfaces of transparent bodies is different for each of them. In general it is greater for bodies which refract light more." But this is as far as he went. A quantitative law—the tangent of the polarizing angle is equal to the refractive index—had to wait until 1813 for David Brewster (1781–1868), a Scottish clergyman turned amateur scientist (and the inventor of the kaleidoscope). Brewster arrived at this law by compiling refractive indices and polarizing angles for hundreds of materials. Because of his contribution, the polarizing angle is often called the Brewster angle (although perhaps the French call it the Malus angle).

The importance of Malus's discovery is that it forced the recognition that polarization is a property of light, not of matter. Malus not only discovered polarization upon reflection, he also coined the term *polarization*. In one of his Queries, Newton had asked, "Have not the Rays of Light several sides endued with several original Properties?" It seems that he envisioned light to be composed of corpuscles (particles) having several sides. These corpuscles are randomly oriented in unpolarized light, but may be aligned upon transmission by a birefringent medium.

In *The Nature of Light*, Vasco Ronchi writes that "by analogy with magnetic bodies Malus suggested that the corpuscles of light had poles, so he called the light which had particles all uniformly oriented, 'polarized light' . . . He found that there was no difficulty in assuming a polarizing action in reflection, because the reflecting surface had the power of orienting the corpuscular bipoles." Thus launched by Malus, the term *polarization* persists today, despite its origin in a faulty analogy between magnetic poles and the asymmetric properties of light.

A Few Misconceptions Dispelled

In some treatments of polarization upon reflection, the Brewster angle gets all the attention, leaving one with the impression that

Figure 4.3

Over a wide range of angles of incidence around the Brewster angle (about 57 degrees), initially unpolarized light acquires a high degree of polarization upon reflection by glass.

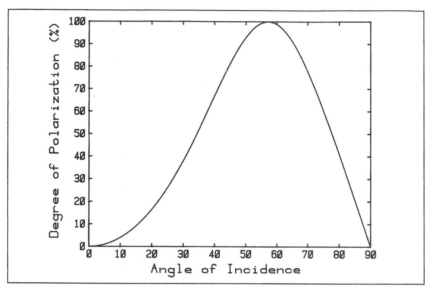

nothing worthy of note occurs at other angles of incidence. Yet the degree of polarization of reflected light varies continuously from 100 percent at the Brewster angle to 0 percent at normal and glancing incidence. This is shown in Figure 4.3.

Note that over a 40-degree interval around the Brewster angle, the degree of polarization of reflected light exceeds 50 percent. This underlies the efficacy of polarizing sunglasses: if they reduced glare only at the Brewster angle, they would be useless except during those fleeting moments when sunlight is incident on the hood of a car or on water exactly at this angle.

The consequences of the curve in Figure 4.3 can be observed readily enough with a polarizing filter and a row of ceiling lights reflected by a smooth floor. I took the two photographs shown in Figure 4.4 in the hall outside my office. Each light is reflected at a different angle, hence the degree of polarization of each reflection varies, decreasing with increased distance. Thus when I rotated the polarizing filter on my camera so that its transmission axis was perpendicular to the floor, the brightness of each image, initially the same, was reduced disproportionately; the

Figure 4.4
Reflections of ceiling lights by a polished floor can be disproportionately reduced in brightness by using a polarizing filter. The top photograph was taken with the transmission axis parallel to the floor, the bottom photograph with the transmission axis perpendicular to the floor. Note that the brightness of the ceiling lights does not change, evidence that the light from them is unpolarized. Reflection by the floor is least when that by the walls is greatest because floor and walls are perpendicular to each other.

Figure 4.5

The degree of polarization of light reflected by chromium rises to nearly 70 percent for an angle of incidence of about 80 degrees.

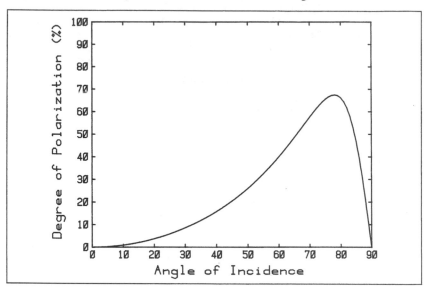

image of the closest light—chosen to be reflecting near the Brewster angle—almost vanished.

The qualifier *transparent* often appears in discussions of polarization upon reflection, beginning with that by Malus. Yet the floor shown in Figure 4.4 is certainly not transparent. Thus transparency cannot be a necessary condition for polarization. What is necessary is that the reflecting surface not be polished metal.

Yet even this statement needs to be qualified. Perfectly conducting metals are indeed worthless as polarizers, but perfection is not of this world. For some metals, the degree of polarization of reflected light is unexpectedly high—if you know where to look. For example, the maximum degree of polarization of visible light reflected by chromium is almost 70 percent (see Figure 4.5), although to observe this you would have to look at light reflected near 10 degrees from glancing. Silver and aluminum give much lower degrees of polarization—but not zero at all angles.

I made another foray into the hall outside my office in search of more grist for my polarization mill. I found it in the form of ashtrays made of what appeared to be stainless steel. When I examined these with a polarizing filter while rotating it, highly oblique reflections waxed and waned in brightness. You can amuse yourself by examining other metals with a polarizing filter and find polarization where—according to those who haven't looked—it isn't supposed to exist.

Back to the Airplane Window

After this long preamble, you may have forgotten the purpose of my previous chapter and this one, which is to explain colored reflections seen in the lenses of my reading glasses while flying over the Pacific Ocean. All the ingredients are now at hand. Indeed, you already may have put them together, a simpler task than assembling them.

Light streaming through the window of my airplane was (partially) polarized because of scattering by atmospheric molecules. This window was a retarder because of stress-induced birefringence. And the second polarizer, the origin of which I left unresolved, was the lens of my glasses. Dangling from my fingers, they inadvertently were oriented so that light from the window was incident on the lens near the Brewster angle.

You don't have to be on a trans-Pacific flight to see the kinds of colors I have been discussing. Several years ago I spent a summer at the Naval Ocean Systems Center in San Diego. One day I came into my building out of the bright sunshine and didn't bother to take off my polarizing sunglasses. As I marched down the hall, I happened to glance into an office. What I saw lying on a table gave me a start: a multicolored plastic triangle. When I took off my glasses, the colors disappeared. Light streaming through the window was partially polarized upon reflection by the table as well as by the triangle, then transmitted by the triangle, a nonuniform retarder. The second polarizer (analyzer) was my sunglasses.

I subsequently have seen colors like those exhibited by the plastic triangle even without the aid of any polarizing filters. For example, a photograph encased in a plastic frame stands on a table in my living room. It is illuminated from above by a lamp. When I look at lamplight reflected first by the frame and then by the underlying smooth table, I see striking colors if my vantage point is just right. The plastic is both a polarizer and retarder; the table is the analyzer.

47

A curious sequel to what I observed over the Pacific occurred in Hawaii, my destination. I had gone there for a meeting on meteorological optics. These meetings often begin with sessions in which each participant tries to excite the envy and admiration of the others with slides of strange and wondrous sights. One man showed a slide of multicolored fractured ice, which he could not explain. He swore that there had not been a polarizing filter on the camera. With the image of the colors I had seen while flying to the meeting still fresh in my mind, I offered the following explanation.

Although skylight is partially polarized, unpolarized sunlight alone is sufficient to give what was observed. Clear ice gives no colors upon reflection, nor does highly fractured ice (e.g., snow). If slightly fractured, however, it would have a few surfaces, some to polarize incident sunlight, others to analyze it. And ice is naturally birefringent. We can therefore imagine the following plausible sequence: Sunlight is incident on one ice surface near the Brewster angle, reflected, then transmitted by a section of birefringent ice, and finally reflected to the observer by another properly oriented ice surface. Thus, colors may be produced with fractured ice in the same way as with airplane windows and other sheets of stressed plastic.

When I returned to Pennsylvania, I wrote to the man with the slide of multicolored fractured ice and asked him for a copy. I also gave him a more detailed explanation of what he had observed, but he never responded. So I must ask you to keep your eyes open for colors in slightly fractured ice during the winter. Multicolored ice must be rare because I have yet to see it, even though I long have had an interest in the colors of ice and snow and keep my eyes open for them.

*5

Light Bulb Climatology

Put out the light, and then put out the light.
WILLIAM SHAKESPEARE: *Othello*

*D*avid Taylor, who not long ago retired as a chemistry professor at the University of Natal in South Africa and is
now living in England, once ended a letter to me with the following question: "While on the subject of car headlights, can
you tell me why they do not go out immediately when you switch
them off, unlike ordinary incandescent lamps? I have asked physicists this, but they usually deny it, which is silly because you
can test it any night in your garage."

This was a question much to my taste, made all the more
delicious by ludicrous denials that what had been observed could
not have been. So I rolled up my sleeves and did a more thorough
analysis of a problem to which I previously had given only per-
functory thought.

Thermal Inertia Is *Not* a Bad Word

Taylor offered a tentative explanation of what he had observed:
"One would suggest there is a greater mass of filament in the
car lamp, which takes longer to cool." I wrote to him that I
agreed with his explanation, although I pointed out that even
ordinary incandescent lamps do not switch off immediately. I
often observe this in my basement, which is illuminated by a
single naked bulb hanging from the ceiling. When I flip off the
switch, the brightness of the bulb decays rapidly, but not so
rapidly that I perceive it as instantaneous.

Incandescent lamps are bright because their filaments are
heated by electric currents to sufficiently high temperatures that
an appreciable fraction of the radiant energy they emit is visible.
When the current in a filament suddenly falls to zero, it cools,

49

mostly by radiant emission, the rate of which is proportional to its surface area.

For simplicity, assume the filament is a long wire, so thin that it can be characterized by a single temperature. The energy of the filament is proportional to its mass, hence volume, and its temperature. Because the filament is radiating energy, its energy will decrease when the current to it is turned off, which is manifested by a decrease in temperature.

Let us consider a time interval during which the rate of emission by the filament is approximately constant. The total emission is the rate of emission times the length of this interval. It is also equal to the energy loss of the filament, which is proportional to its temperature decrease and volume. This yields the result that the temperature decrease is proportional to the time interval, the rate of radiant emission, and the ratio of the surface area of the filament to its volume. This ratio is inversely proportional to its diameter but independent of its length because area and volume depend on length in the same way.

Let us turn this around and ask not what the temperature change is in a given time, but rather how much time it takes the temperature to decrease by a given fraction. This time, which I shall refer to as the *thermal inertia*, is directly proportional to the diameter of the filament and inversely proportional to its rate of emission.

Thermal inertia continually increases during cooling because the rate of emission is strongly dependent on temperature. So there is not a single thermal inertia, but an infinitude of inertias, each one greater than its predecessor because of a continually decreasing rate of emission. Nevertheless, we can talk about the average thermal inertia of a filament during cooling, which is proportional to its initial value.

We observe the temperature of incandescent filaments not directly, but rather indirectly by way of their *luminance* (strictly speaking, their *brightness*; see Chapter 15 for the distinction between luminance and brightness). Because the luminance of an incandescent body depends so strongly on its temperature, we soon cannot see the radiation emitted by a cooling filament. You can observe this with an electric stove. When the electric current to one of its filaments is turned off, it quickly cools and is no longer luminous, although you must wait much longer before you can touch it.

Always eager to avoid controversy, I hesitate to coin the term *luminance inertia*, even though it is the consequence of this inertia, rather than the thermal inertia, we see when we turn off an incandescent lamp (the two are proportional, of course). It seems that *thermal inertia* was offensive to the late Konrad Buettner, so much so that he published a brief letter in *Applied Optics* (Vol. 8, 1969, p. 212) entitled "Thermal Inertia Is a Bad Word." Yet he failed to convince me that thermal inertia is a "very misleading term." To the contrary, it seems well suited to convey the idea that things take time to cool (or to heat).

Yves Le Grand, in his book *Light, Colour, and Vision* (also cited in Chapter 9), uses *thermal inertia* (p. 307) without flinching. So great is his confidence that this term is charged with meaning, he doesn't even bother to define it. And at a recent gathering of friends, the term *thermal inertia* entered into conversation without anyone blanching or hustling the children out of earshot.

Invoking thermal inertia can be just an imprecise way of acknowledging that it takes time for objects to cool or to heat, and that this time increases with their size. I have tried to make *thermal inertia* more precise by defining it as a time. *Thermal inertia* so defined is consistent with what is usually meant by *inertia*. According to one dictionary I consulted, *inertia* is "resistance to motion or change." Resistance to motion has a quantitative embodiment in the concept of mass, but we could define it as a time—the time it takes a body to undergo a given fractional change in velocity when acted upon by a given force. This time is directly proportional to both the size of the body (i.e., its mass) and its velocity, and inversely proportional to the force acting to change its velocity. Similarly, the thermal inertia of a filament, defined as a time, is directly proportional to its size (i.e., diameter) and inversely proportional to its radiant emission rate, which is what causes it to change its temperature.

To be fair to Buettner, now that I have had a bit of fun with the title of his letter, I doubt that he would disapprove of the way I use *thermal inertia*. He inveighed against the use of this term in a somewhat different context.

Temperature and Luminance Variation

With my pencil still sharp, I decided to go beyond Taylor's question and consider the temperature and luminance variations of light bulbs while they are turned on. Among other things, I addressed the question of why the light from them appears constant even though it is driven by a varying current.

In the United States, household electric current oscillates at 60 Hz (cycles per second). At one instant the current is a maximum and flowing in one direction; 1/120th of a second later, it is a maximum in the opposite direction; in another 1/120th of a second it again peaks in the direction it had when the cycle began. Heating by the current, however, does not depend on its direction. The rate of heating of a filament is proportional to the square of the current in it. Thus a light bulb filament undergoes periodic heating at a frequency of 120 Hz.

At one instant the heating is a maximum; 1/240th of a second later, it falls to zero; 1/240th of a second later, it again peaks. And so on ad infinitum—or at least until the switch is turned off or the electricity bill isn't paid. Because of thermal inertia, the temperature of the filament cannot follow the comparatively high-frequency heating oscillations. So although the heating rate oscillates rapidly between zero and its maximum value, the temperature oscillations are more subdued (see Figure 5.1). The difference between the maximum and minimum filament temperatures relative to the average is proportional to the ratio of the heating period (1/120th of a second) to the thermal inertia. On the basis of many observations of the waning luminance of light bulbs when they are turned off in otherwise dark rooms, I estimate that their thermal inertia is about one second or less. It cannot be as much as five seconds or I could time the luminance decay with a stopwatch. And it probably cannot be much less than 1/10th of a second or I would perceive the decay as instantaneous. The precise value of the thermal inertia is less important than is the fact that it is appreciably greater than the heating period.

The luminance of a heated filament is not linearly proportional to its temperature, but if the temperature variation is small, the difference between the maximum and minimum luminance relative to the average is, like the relative temperature difference,

Figure 5.1
The rate of heating of a light bulb filament by the electrical current in it oscillates between zero and its peak value (dashed curve). Because of the thermal inertia of the filament, the departures of its temperature from the average value (solid curve) are more subdued. Moreover, the temperature and heating rates are not in phase, the heating rate peaking before the temperature does.

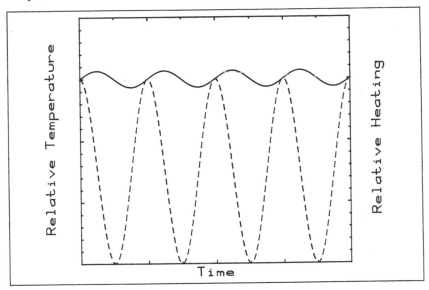

proportional to the ratio of the heating period to the thermal inertia. So the luminance variations of incandescent lamps are 1 percent or less. This is one reason why we perceive their luminance to be continuous: its variations are below our threshold of detection. I shall defer discussing the second reason until the following section.

The temperature of the filament does not exhibit the same relative variation in amplitude as the heating rate, and the two are out of phase. That is, they do not reach their peak values simultaneously, the temperature peaking after the heating rate (Figure 5.1). This phase lag, like the amplitude variation, depends on the ratio of the thermal inertia to the heating period. If the thermal inertia is small compared with the heating period, the temperature and heating rate are in phase. The greater this ratio, the greater the phase lag. Its maximum value is 90 degrees: the temperature can peak at most one-fourth of a cycle later than the heating rate.

After I completed my analysis of the variations in amplitude and phase of the temperature and luminance of incandescent lamps, it occurred to me that Irving Langmuir must have considered this problem. And sure enough, I found a paper by him entitled "The Flicker of Incandescent Lamps on Alternating Current Circuits and Stroboscopic Effects," originally published in *General Electric Review* (March 1914) and included in his *Collected Works*. Langmuir cited earlier work by Orso Mario Corbino, which I dug out of the archives. I was pleased to see that what I had done was not much different from what they did.

Langmuir, who died in 1957, was one of the most distinguished scientists of his generation. He spent almost his entire career with General Electric, hence his interest in lamps. No industrial scientist has to my knowledge surpassed him. His *Collected Works* run to 12 volumes, impressive evidence of his contributions to many fields. He is perhaps best known for his work in the physics and chemistry of surfaces, for which he was awarded the 1932 Nobel Prize in chemistry. Within the past few years, the American Chemical Society has published *Langmuir*, a journal devoted to surface and colloid science. Thus Langmuir has the rare honor of being one of the few scientists after whom a scientific journal is named.

Langmuir was also a pioneer in plasma (a term he coined) physics, the physics of ionized gases. The last 18 years of his scientific career were devoted to atmospheric science, his contributions to which occupy two volumes of his *Collected Works*.

Observations of Luminance Variations of Lamps

I alluded to two reasons why light from incandescent lamps appears constant but discussed only one (thermal inertia greatly dampens temperature variations, hence luminance variations). The second reason is that when we look at a source of light consisting of a series of brief flashes at regular intervals, we do not detect the instantaneous luminance but rather its average value if the frequency of the flashes lies above what is called the *critical flicker frequency* or *fusion frequency*. This is the Talbot-Plateau law. Because of this we would perceive the luminance of incandescent lamps as constant even if their thermal inertia were as small as that of fluorescent lamps.

The value of the fusion frequency depends on many factors, one of the most important of which is the (average) luminance. Le Grand states that the value of the fusion frequency may "fall to 3 or 4 per second when the luminance is very low and may exceed 100 when it is at its highest; this explains why fluorescent lamps, which do not possess the thermal inertia of filament lamps, may sometimes flicker under certain viewing conditions, in spite of the fact that their frequency is . . . 120 in the United States of America."

It is possible to dodge the Talbot-Plateau law by moving your head rapidly while observing a distant flashing light. Because I live in the woods and do not own a television, I have to find ways to amuse myself after the sun goes down. One way is to observe distant lights against an otherwise dark background through binoculars while moving them rapidly. By means of this curious maneuver, you can readily distinguish between gas-discharge (e.g., fluorescent) and incandescent lamps.

Move the binoculars rapidly in a kind of circular motion so that the image of the lamp traces out a luminous curve on your retina. Depending on the nature of the light source, this curve will be either continuous or intermittent. The luminance of the gas-discharge lamp responds very rapidly to variations in the alternating current. Because of its much greater inertia, however, the luminance of the incandescent lamp is sensibly constant. Thus when you look through moving binoculars at such a lamp, the luminous curve on your retina is continuous. But when you look at a gas-discharge lamp, this curve consists of alternating dark and bright segments. You are resolving in space what you cannot resolve in time: the lamp is on (i.e., maximum current) when its image is on one part of your retina, but during the time it took you to move the image to another part, the lamp has gone off (i.e., the current has dropped to zero). When you stare fixedly at the lamp, you cannot perceive this temporal variation of luminance because its frequency is greater than the fusion frequency.

Until a few years ago, I had never thought of photographing what I have observed through rotating binoculars. I must have thought it impossible. But one evening, in a bold and reckless mood, I decided to give it a try. I loaded my camera with the fastest film I could find (ASA 1600), snapped on a telephoto lens,

Figure 5.2

The top photograph shows the continuous luminous curve obtained when a camera was rotated for about a second with its shutter open while pointed at an incandescent light. The camera was pointed at a mercury vapor lamp for the bottom photograph; the total exposure time, about one-half second, is evident from the number of luminous segments traced out at a frequency of 120 Hz.

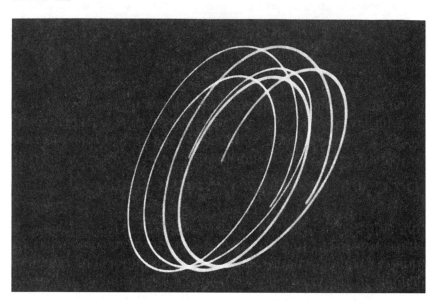

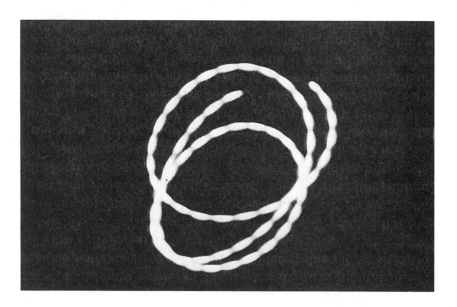

Figure 5.3
The points connected by solid lines are hourly average temperatures on a mostly clear spring day in State College, Pennsylvania. The points connected by dashed lines show the corresponding measured solar radiation (in arbitrary units); the gap in the radiation data resulted from intermittent clouds. Note that the temperature peaks almost four hours after the solar radiation.

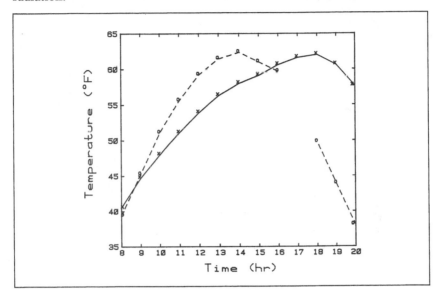

and waited for the night to deepen. We have two types of outdoor lights on our property: a mercury vapor, dusk-to-dawn light on a pole in front of our house and incandescent lights on the porch. I got far away from them, aimed the camera at each in turn, opened the shutter, wiggled the camera for about a second, closed the shutter, then advanced the film for another try. The results are shown in Figure 5.2. I captured even the intermittent light of a gas-discharge lamp a few miles away in the valley below.

Light Bulb Climatology

After I had done my analysis of the temperature variations of incandescent lamps, it occurred to me that there are parallels between such lamps and the atmosphere. In particular, the phase difference between the heating current and the filament temperature has its parallels in the lags between solar radiation and temperature, both diurnal and annual. Figure 5.3 shows, for example, how the temperature varied from sunrise to sunset on a

Figure 5.4
The solid curve shows the solar radiation (arbitrary units) at the top of the atmosphere in State College. The horizontal lines are average daily temperatures between 1950 and 1981. Note that the temperature peaks about a month after the solar radiation.

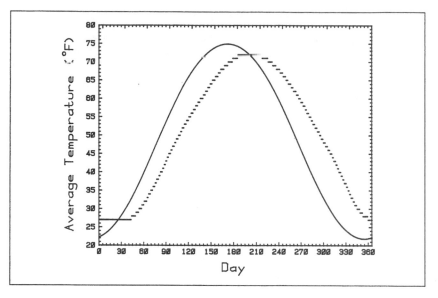

sunny April day in State College, Pennsylvania. Also shown is solar radiation (measured at the ground), the variation of which, to a large extent, determines the temperature variation. Air temperature is analogous to the temperature of a filament; solar radiation is analogous to the electric current heating it. Note that the peak temperature lags the peak solar radiation by almost four hours. Of course, even casual observers know that the hottest time of the day is usually not solar noon. It is not so obvious that the month with the most sunshine is unlikely to be the hottest month. This is evident in Figure 5.4, in which average daily temperatures from 1950 to 1981 in State College are plotted together with the solar radiation. The warmest month is July, whereas the month with the most solar radiation (at the top of the atmosphere) is June, yet another example of the consequences of thermal inertia.

The variation of filament temperature and its phase lag are related: the greater the phase lag, the more subdued the variations in temperature. In keeping with the analogy between light bulbs and the atmosphere, there ought to be a similar relation between atmospheric temperature variations and phase lags. When I mentioned this to Cliff Dungey, he told me that the temperature in Moscow, which has a continental climate, exhibits large variations in amplitude and a small phase lag, whereas the temperature in Vladivostok, which has a maritime climate, exhibits smaller amplitude variations and a larger phase lag. He also pointed out that annual soil temperatures exhibit the same kind of inverse relation between amplitude and phase. Good examples of this are given in a paper by Singer and Brown in *Transactions of the American Geophysical Union* (Vol. 37, 1956, p. 743). Soil temperatures they measured a few feet below the surface varied about 22°C (40°F) over the year and peaked in mid-August, whereas at about ten feet the annual temperature swing was halved and the peak occurred a month later. Deeper still, the temperature hardly varied at all, and its peak occurred even later.

Caveats and Conclusions

No explanation can be considered complete until competing ones have been weighed and found wanting. Although I did not say so previously, the thermal inertia of filaments also depends on the material out of which they are made. In attributing Taylor's observation to a greater filament diameter for headlights than for light bulbs, I implicitly assumed that all filaments are made of similar materials. This assumption seems reasonable, although I have not verified it. Nor have I made a tour of junkyards, armed with a micrometer, to measure the diameter of headlight filaments. Despite these loose ends, I am still of the opinion that any noticeable differences in the thermal inertias of incandescent lamps result mostly from different sizes, and possible shapes, of their filaments.

Automobile headlights operate under voltages much lower than household voltages. Their filaments must therefore have a smaller electrical resistance to yield even the same, let alone greater, power outputs as light bulbs. All else being equal (i.e., composition and length), the resistance of a filament decreases as its diameter increases, which supports the hypothesis that filaments in automobile headlights are thicker than those in light bulbs. I also have noticed that the inertia of high-wattage bulbs is greater than that of low-wattage ones. Again, this is consistent with a thicker filament for the high-wattage bulbs: for a fixed voltage, the power increases as the square of the filament diameter.

What we observe when a light is switched off is the time it takes the temperature of a filament to decrease to where we cannot see the radiation emitted by it. This time will be longer the greater the temperature of the filament at the instant cooling begins. For a fixed voltage, the current in the filament will increase with power (wattage), and the greater the current it carries, the greater its temperature. Thus it takes more time for a filament to become invisible when its power is higher. Because both filament power and thermal inertia increase with thickness, the result is a greater luminance inertia. Temperature (i.e., power) is less important than thickness in determining this inertia, however, because of the partially compensating effect of a higher rate of average emission during cooling for greater initial temperatures.

Nothing occurs in an instant. Everything takes time. There are actually three inertias involved when one throws a switch: the electrical inertia of the circuit (currents do not decay to zero instantaneously), the thermal inertia of the filament, and the physiological inertia of the eye-brain system (i.e., after-images, the persistence of a response after the stimulus has been removed).

To at least estimate the magnitude of the first and last of these inertias, I observed the luminance decay of a fluorescent lamp (which has negligible thermal inertia) as I repeatedly switched it off in an otherwise dark room. I looked through a paper tube at the lamp so that its angular width would appear to be about the same as that of a light bulb. To my eyes, the fluorescent lamp turned off much faster than does a 150-watt light bulb. So my guess is that electrical and physiological inertias do not contribute appreciably to the total luminance inertia of light bulbs.

I hope that, after having read this chapter, you will no longer take for granted the simple act of turning off a light switch. A light bulb is a world unto itself, and its behavior parallels that of the larger world around it.

Highway Mirages

O! swear not by the moon, the inconstant moon
WILLIAM SHAKESPEARE: *Romeo and Juliet*

*E*very summer for nine years, beginning in 1980, my wife and I made the long trek from Pennsylvania to Arizona, or New Mexico, or California. In the summer of 1989 we stayed home, perhaps a signal of decline into a more sedentary way of life. But the nomad in me has not yet been fully tamed, and, chained to my desk, I sometimes reflect wistfully on past scenes that have caught my eye. Over some of the more dreary sections of the country—unnamed to spare myself letters from irate inhabitants of the hinterlands—the roadside scenery has not been sufficiently interesting to relieve my boredom, and I have had to turn to the road itself for diversion.

For hours on end I have watched with fascination mirages such as that shown in Figure 6.1. Sometimes I have photographed them through a telephoto lens, or just watched them through the lens or through binoculars—while my wife was driving, of course. Although you can observe highway mirages with the unaided eye, modest magnification greatly enhances your enjoyment of them.

What is it that causes highway mirages, and what do they reveal to the discerning eye?

A Mirage Is Not an Illusion

I shudder when I see or hear mirages referred to as optical illusions. Even rainbows are sometimes so maligned, as well as all the splendors of the sky caused by ice crystals. Yet these optical phenomena are no more illusions than are images in a mirror. Light really does come from a mirror, there just isn't a flesh-and-

Figure 6.1
The distant white car is accompanied by its inverted image, a mirage resulting from refraction by air overlying a warmer surface. Note also that the grass in the center strip appears to be encroaching on the highway, yet another mirage.

blood person trapped inside of it (which my dog learned even before he was housebroken).

When the moon (or sun) is low on the horizon, it is perceived to be much larger than when overhead. This is a true illusion, which even today, almost a thousand years after its psychological origin was recognized, is still sometimes wrongly attributed to atmospheric refraction. The moon illusion results from refraction by the mind, mirages from refraction by the atmosphere. When images formed by the refracting atmosphere depart markedly from what they would be in its absence, they are called mirages.

Refractive Index Gradients

The atmosphere is optically nonuniform: its refractive index varies from point to point, especially along vertical lines. Within a few meters of the ground, vertical refractive index variations are almost entirely the result of temperature variations.

The refractive index of air depends on the *number* density of the air's molecules—the number of molecules packed into a given volume. This is not surprising since it is molecules that do the refracting. Number is italicized to dispel the misconception that the refractive index of a material depends on its *mass* density. The concept of mass is in no fundamental way linked with that of refraction. Points on a plot of refractive index versus mass density for many liquids and solids would not lie on a smooth curve but would be scattered randomly as if made by firing a shotgun at a piece of paper. It is true, however, that the refractive index of a *gas* at normal temperatures and pressures is directly proportional to its molecular number density.

The number density of a gas such as air depends on its pressure and temperature. If pressure is held constant, the number density increases with decreasing temperature; if temperature is held constant, the number density decreases with decreasing pressure. Because both pressure and temperature change with height, both contribute to changes in the refractive index of air.

Pressure decreases with height, from the surface of the earth to the outer fringes of the atmosphere. With this pressure decrease acting alone, the refractive index of air would decrease steadily with height. A good average value for the temperature gradient or lapse rate (its rate of decrease with height) in the lowest ten to fifteen kilometers of the free atmosphere—that part well above the surface—is about 6.5°C/km. For such a gradient, the refractive index of air would increase with height but at a rate about five times smaller than it decreases because of decreasing pressure. Thus the refractive index of air steadily decreases with height under average conditions in the free atmosphere because the consequence of decreasing pressure outweighs that of decreasing temperature.

Near the ground, however, conditions depart markedly from those higher up: the temperature gradient can be ten, a hundred, even a thousand times the free-atmosphere value. Temperature can even increase with height. Mirages are seen along lines of sight that lie near the ground; thus the existence and form of mirages are determined mostly by temperature gradients.

Before considering the consequences of atmospheric refractive index gradients, I must dispose of the notion that water vapor plays an essential role in the formation of mirages. This misconception dates to antiquity and persists today, evidence that no misconception ever dies. Water vapor does indeed contribute to

63

air's refractive index, hence its gradient, but this vapor makes up only a few percent of the molecules in the atmosphere. And at visible wavelengths, the contribution of each water molecule to the refractive index of moist air is slightly less than that of each nitrogen or oxygen molecule.

Ray Tracing in a Nonuniform Atmosphere

In an optically uniform atmosphere, one with the same refractive index everywhere, paths of light rays are straight lines. In a nonuniform atmosphere, such paths can be curved. A hypothetical refractive index profile near the ground is shown in Figure 6.2. Although I confected this profile, it is not unlike temperature profiles measured over heated surfaces. Within a few meters of the ground, this profile corresponds to a lapse rate about 500 times the free-atmosphere value. The refractive index shown is for a wavelength near the middle of the visible spectrum, where the human eye is most sensitive. The atmosphere is *dispersive*: its refractive index varies with wavelength. Although this can give rise to observable phenomena such as the green flash, the dis-

Figure 6.2
This hypothetical refractive index profile is not unlike those that occur over strongly heated surfaces. The horizontal axis shows the departure of the refractive index of air from unity magnified by 1,000.

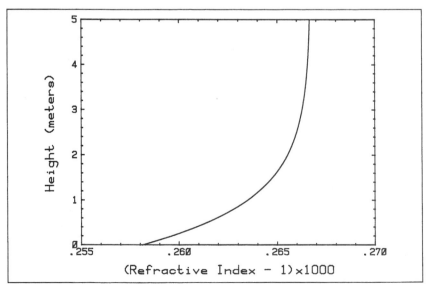

persiveness of the atmosphere does not affect the appearance of mirages. We could base our reasoning on refractive index profiles at any visible wavelength with essentially identical conclusions.

For many purposes, we may take the refractive index of air to be exactly one. Understanding mirages, however, is not one of these purposes. Although the departure of the refractive index of air from unity is only a tiny fraction of one percent, it is this small departure—or rather its variation with height above the ground—that causes mirages.

The trajectories of a few light rays in an atmosphere with the refractive index profile in Figure 6.2 is shown in Figure 6.3. Note the extreme distortion of scale. Pictorial representations of mirages are necessarily distorted. Vertical distances must be greatly exaggerated. If they were not, ray trajectories would appear straight. Scale exaggeration is done to help you understand mirages, but it should not mislead you about their angular size. Mirages subtend only about a degree. Lest you think that this isn't much, recall that the sun and moon subtend one-half of one degree.

Figure 6.3
These ray paths are those appropriate to air near the ground with the refractive index profile shown in Figure 6.2. Although the object point *A* is above the observer at *B*, he interprets *A* as lying below him because of the direction of the ray from *A* at his eye. When he moves to *C*, he sees two images of *A*, one erect and one inverted.

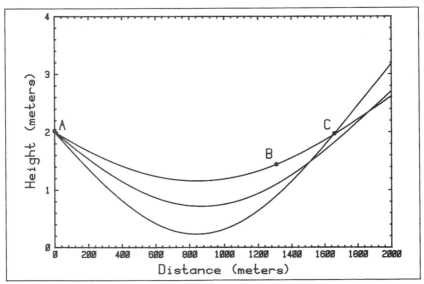

How we interpret images depends on the directions of rays at our eye, not at their source or at intermediate points. If the atmosphere were uniform, a ray from *A* to *B* would be a straight line. An observer at *B* would see *A* in its true position, slightly above (in angle) the horizontal. But in a nonuniform atmosphere, the ray from *A* that enters the observer's eye at *B* is seen *below* the horizontal. Thus the observer interprets *A* as lying below him. This is called an *inferior* mirage, which does not define its social status but merely indicates that images are displaced downwards from the positions they would occupy without refraction. Inferior mirages are contrasted with *superior* mirages, seen when temperature increases with height. Because an asphalt surface bathed in sunlight is hotter than the air above it, highway mirages are usually inferior mirages.

Note in the profile in Figure 6.2 that the gradient of the refractive index, its rate of change with height, is greatest at the ground and decreases steadily with height. Above about four meters, the refractive index is nearly constant. The angular displacement of an image point below its object point is proportional to the refractive index gradient: no gradient means no displacement. Unless the gradient is constant, unlikely though not impossible, images of the lowest points on an object will be displaced downward more than images of points higher on the object. As a consequence, the image of an object in an inferior mirage is not only displaced, it is usually magnified as well. But this does not exhaust the possibile distortions. Indeed, we have yet to explore the most striking ones.

Multiple Images

An observer at *C* (Figure 6.3) receives two rays from the same point, hence sees a double image. You may have seen multiple images in reflected light many times without recognizing them as such. For example, the image of the sun low over the sea (Figure 6.4) or a lake ruffled by wind is not circular, as it would be if the water were perfectly flat, but appears as a long narrow band. This band is a great many images of the sun formed by the wavy surface, which is shown schematically in Figure 6.5. A single object point is the source of rays that are reflected by different waves and therefore arrive at the observer's eye from different directions.

Figure 6.4
This narrow band of light is a series of multiple images of the setting sun reflected by different waves. If the sea were flat, the reflected image of the sun would be a disc.

Figure 6.5
Light rays from a single point are reflected by different waves in different directions, thereby giving multiple images of the point.

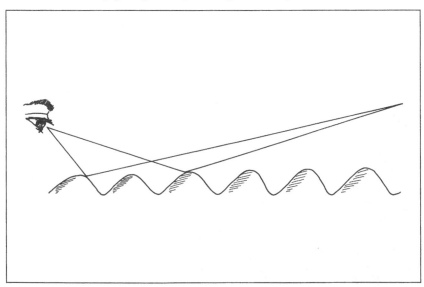

It is possible for multiple images to occur in refracted light as well as in reflected light. Although not obvious in the ray diagram (6.3), one image of point *A* seen by the observer at *C* is inverted and the other erect, the inverted image lying below the erect one. Both images deserve the appellation *mirage* since both are different from what would be seen in the absence of a temperature gradient; the inverted image is merely more obviously different.

The change in the mirage seen by an observer who moves from *B* to *C* is striking, just one example of how a mirage depends strongly on the position of its observer relative to its source. As you watch cars pass you on the highway, which I frequently do (our personal license plate advises drivers to PASSUS), they often grow a second image as they recede into the distance (Figure 6.6). You also may notice that such cars often appear to be driving over a wet patch on the pavement, which has mysteriously dried by the time you reach it. The impression of wet pavement ahead is made more compelling by the inverted image of a car, just

Figure 6.6
The nearby cars in the left lane are seen without noticeable distortion. But the distant truck in the right lane is seen as two images, one erect the other inverted. Note the second image of the truck's white mud flaps. Inverted images of the farthest cars in the left lane also can be seen, although they are not so distinct as the images of the mud flaps.

Figure 6.7
What appears to be water on the pavement around the truck is an inverted image of the sky.

what would be seen if the car were at the edge of a pond. It is this apparent water on highways and deserts that has given mirages their undeserved reputation as illusions.

Sources of nonexistent water on highways are not hard to find. The truck shown in Figure 6.7 appears to be driving over a patch of water on the pavement. This patch is simply the inverted image of the horizon sky. Once we learn this, we no longer are fooled—if we ever were. This is much like the experience of my dog when he was a puppy and first faced his image in a mirror. He jumped back, startled; barked at his twin; approached it cautiously, sniffed it; then spurned it as if to say, It's only an image. Ever since, he has been indifferent to what he sees as he passes mirrors. He has learned that other dogs don't live inside of them. Neither he nor we find mirror images illusory. We are not fooled by them. They have their origins outside of our minds. Contrast mirror images (reflection images) and mirages (refraction images) with a true illusion, the moon illusion.

We learn early in our lives that the images we see in mirrors are not objects. We also can learn that the splashes of apparent water seen on highways are inverted images of the sky. We do so by tracing rays, which lead us to the physical origins of these splashes. But we cannot trace rays that will lead us to the origins of the moon illusion. Whenever we see the moon on the horizon, we interpret it as being larger than it really is, and no amount of learning can disabuse us of this. Intellectually, we can accept that the moon is not objectively larger, but it will always appear larger to us. An enlarged moon is a creation of the mind; a mirage is a creation of the atmosphere.

Although I began this chapter with tales of summertime mirages, you don't have to wait until summer; you can see highway mirages all year round, even in winter when the sides of the road are piled high with snow. High temperatures do not give rise to mirages; high temperature gradients—the rates at which temperature changes with height—do.

*7

The Greenhouse Effect Revisited

By the influence of the increasing percentage of carbonic acid in the atmosphere, we may hope to enjoy ages with more equable and better climates, especially as regards the colder regions of the earth
SVANTE ARRHENIUS: *Worlds in the Making* (1908)

Several years ago, I wrote a series of articles in which I rashly predicted that the greenhouse effect might give way to nuclear winter as the global catastrophe most favored by headline writers. Subsequent events have shown otherwise. Nuclear winter is fading into obscurity, whereas only a hermit snug in a cave is spared news of the greenhouse effect. Newspapers hardly let an issue go by without warnings of impending global warming. At breakfast, I learn that Cleveland will become an oasis, whereas the verdant and fertile fields of Iowa will be transformed into an American Empty Quarter.

Concern accelerated during the drought of 1988, taken by some to be a portent of the grim future awaiting us. As the drought stretched on, a trickle of articles on the greenhouse effect turned into a stream, then a raging river, finally a flood of biblical magnitude. It may have crested by now. I even sense a waning, but I have been wrong before.

Although a failure as a prognosticator, I may have been one of the first to observe a new malady, the Greenhouse Effect Anxiety Syndrome (GREAS). Not long ago, a man called my department. He wanted to talk to "a meteorologist," and so was passed on to me by a receptionist with a malicious sense of humor. He did not ask any questions of substance; he just wanted to talk to

71

a sympathetic listener, to vent his anxiety about the impending doom he had read about in a newspaper.

Many of the writers who are giving night sweats to people like him don't bother to learn much about their subject or to inject any freshness into their writing. They rely on stock formulas, reminding us ad nauseam that "the earth radiates heat, this heat is trapped by our atmosphere, which acts like a blanket." These are shaky metaphors at best, shopworn from overuse. We can understand the greenhouse effect without invoking notions of radiation trapping and atmospheric blankets. With this end in mind, I offer the following experiments and meditations on them as a small step toward grasping some of the subtleties of the greenhouse effect.

Cooling Curves

For my experiments, I used a 250 ml glass flask filled with water heated to about 75°C (167°F). Every 10 or 15 minutes, I measured the temperature of the water, thereby generating a cooling curve. According to Newton's law of cooling, the difference between the temperature of a cooling body and that of its surroundings decreases exponentially with time. Thus a plot of temperature on a logarithmic scale versus time on a linear one should yield a straight line. Its slope is inversely proportional to the *cooling time*, which I define as the time required for the temperature difference to drop to half its initial value. The larger the cooling time, the more slowly the flask cools.

Figure 7.1 shows two cooling curves. For a bare flask, the cooling time is about 50 minutes. When the flask is wrapped in aluminum foil, the cooling time increases to 78 minutes. To explain why requires some backtracking.

All bodies emit electromagnetic radiation of all wavelengths at all times. Yet at normal temperatures, say between about 250 and 330°K, the emission spectrum of most terrestrial objects peaks at an infrared wavelength of around 10 μm; shortward of 4 μm and longward of 20 μm they emit much less, but not an unmeasurable amount. Emission is merely small enough that we need not consider radiation outside this range if our interests are limited to energy transfer.

Water in the flask cooled because of energy exchange between it and its surroundings. For the moment, we are concerned

Figure 7.1

These curves were obtained by allowing a 250 ml flask filled with heated water to cool in still air. The temperature difference between the water and that of its surroundings is plotted on a logarithmic scale. Data points indicated by crosses are for a bare flask; those indicated by circles are for a flask wrapped in aluminum foil. Because the infrared emissivity of aluminum is less than that of glass, the foil-wrapped flask cools more slowly.

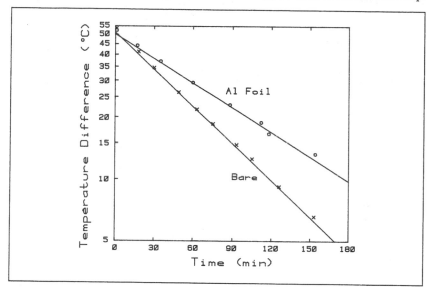

only with the radiative part of this exchange: the flask radiates to its surroundings and they radiate to it. The amount radiated by each depends on their temperatures. But this is not all. Radiant emission also depends on the radiator's composition.

The flask experiments suggest that emission by the foil was less than that by the glass. The foil is so thin and its thermal conductivity so large that it provides little insulation in the sense usually meant—suppression of energy transfer by conduction. But aluminum foil does provide insulation of a different sort, radiative insulation, which brings us to the concept of *emissivity.*

Emissivity

A blackbody is one that absorbs completely all radiation incident on it regardless of wavelength, direction of incidence, and even state of polarization. The emission spectrum of such a hypothetical body depends only on its temperature. At any wavelength, radiant emission by a real body is less than that by a blackbody, both at the same temperature. One way to understand why is to recognize that emission is the reverse of absorption. Suppose that radiation of a given wavelength is incident at a given direction on a body. We may look upon this radiation as a stream of incoming photons, some of which are absorbed by the body. We further imagine that we can film this process (of course, we cannot literally photograph photon paths) and obtain a motion picture of photon absorption. When we run the film backwards, absorbed photons become emitted ones.

The fraction of the incident radiation absorbed by a body is called its absorptivity. Its emissivity is its emission rate relative to that of a blackbody at the same temperature. Because of the film-reversing arguments I made previously, emissivity equals absorptivity. This is called Kirchhoff's law and is strictly valid only for a specified wavelength, direction, and polarization of the radiation. Equality of absorptivity and emissivity is necessary for thermal equilibrium. For example, a body placed in an opaque cavity is bathed in blackbody (or equilibrium) radiation. When the body's temperature becomes constant and equal to that of the cavity walls, the rate of emission by it must equal the rate of absorption.

All radiation incident on a body is not necessarily absorbed; some is reflected and some transmitted. The fraction reflected is the reflectivity; that transmitted is the transmissivity. Because incident radiation can be only absorbed, reflected, or transmitted, the sum of the absorptivity, reflectivity, and transmissivity must be one. If the object is opaque, its transmissivity is negligible, so its absorptivity is one minus its reflectivity. From Kirchhoff's law, the absorptivity is equal to the emissivity, hence that of an opaque, highly reflecting body is low.

Although our eyes cannot tell us, aluminum foil is highly reflecting not only at visible wavelengths but well down into the infrared and beyond. It also is opaque in the infrared. Thus we conclude that it has a low infrared emissivity.

Radiative exchange between the flask and its surroundings depends on the emissivities of both. To good approximation, the surroundings in which I did my experiments are black (i.e., have an emissivity close to one). The net radiative exchange between the flask and its surroundings is proportional to the difference between the fourth power of the absolute temperature of the

74

flask and that of its surroundings, where the proportionality factor contains the emissivity of the flask (strictly, its emissivity averaged over its emission spectrum). Thus, the lower the emissivity of the flask, the less the radiative exchange with its surroundings and the more slowly the water in it cools. A lower infrared emissivity for aluminum than for glass is at least consistent with the different measured cooling times.

Now some of you might say, Bosh! The aluminum foil keeps the flask warmer because it reflects the infrared radiation emitted by the glass. To test this hypothesis, I painted one side of some aluminum foil black, then wrapped it around the flask with the black side inward. The cooling time was only a few percent lower than that for the flask wrapped with unpainted foil, an insignificant difference given the accuracy and reproducibility of my measurements.

The reason for the irrelevance of the radiative characteristics of the foil's inner surface is that the foil and the flask are at nearly the same temperature. The net radiative exchange between the glass and the inner surface of the foil depends on the temperature difference between them. Because this difference is small, so is the net radiative exchange regardless of the emissivities of glass and foil.

In refuting that reflection by the foil is what causes a flask wrapped with it to cool more slowly than a bare one, I was deceptive: I said that I painted the inside of the foil black. Black to my eyes, that is, but neither you nor I see in the infrared. Our eyes are powerless to tell us if black paint at visible wavelengths is also black (i.e., has an emissivity close to one) in the infrared. Indeed, it takes a conscious act of will to convince oneself that an object perceived as black to visible radiation may not be black (i.e., have a low reflectivity) to infrared radiation, and conversely. One way to drive this home is with a simple experiment.

I obtained two cooling curves, one for a flask wrapped in foil painted black and one for a flask wrapped in foil painted white (see Figure 7.2). The cooling time for the white flask was 52.6 minutes, that for the black flask was 50.8 minutes, and that for the bare flask was 50 minutes. Thus the infrared radiative characteristics of glass, black paint, and white paint are nearly identical.

Figure 7.2
To our eyes, the flask wrapped in unpainted aluminum foil (top right) looks more like the flask wrapped with foil painted white (top left) than like the flask wrapped with foil painted black (bottom). At infrared wavelengths, however, the white and black flasks are nearly identical. Both white and black paint films are black in the infrared (i.e., have low reflectivities), whereas aluminum is highly reflecting to infrared as well as to visible radiation.

Engineering students are taught that covering a metallic furnace pipe with insulation can increase rather than decrease the rate of energy transfer from it to the surroundings. This was discussed as long ago as 1933 in *American Journal of Physics* by William Schreiver, who cited experiments done even earlier (1920) at the University of Illinois. To make his point, Schreiver did the same kind of cooling-curve experiments as I have done.

If you were to ask anyone how to decrease the rate of cooling of an object, you would likely be told to wrap it with insulation (wool, for example). This suggestion is so much in accord with common sense that to hint that it might not yield the desired results borders on heresy. Ever the heretic, I wrapped a swatch of thick wool cloth around my foil-covered flask filled with hot water, then measured its cooling time. The cooling time increased by less than 2 percent. Although wool giveth, it also taketh away: it is a much better conductive insulator than aluminum foil, but a much worse radiative insulator. So the net effect of dressing the flask in a warm wool coat is almost nil.

Lest I mislead you, I must point out that the radiative characteristics of a cooling (or warming) body become immaterial when other modes of energy transfer (e.g., convection) predominate. To demonstrate this, I placed a foil-covered flask about one-half meter from a low-speed fan tilted downward. I estimate the wind speed to have been a few knots. The cooling time was 23 minutes. Then I painted the foil black and repeated the experiment; the cooling time was similar, about 25 minutes. In still air, however, the two cooling times differed by more than 50 percent.

These results also demonstrate why *the greenhouse effect* when applied to the atmosphere is somewhat of a misnomer. Greenhouses exchange energy with their surroundings both radiatively and convectively. In even a light wind, convection dominates and the emissivity of the glass becomes immaterial.

Do Good Absorbers Go to Heaven?

Kirchhoff's law is often transformed into a slogan: "A good absorber is a good emitter." Whenever I see or hear this, I imagine patting an absorber on the head and murmuring approvingly, "You're a good little absorber." I was pleased to discover that my distaste for this kind of anthropomorphism was shared by John Henry Poynting, whose name—if not career—is well known to all students of electromagnetic theory, having been enshrined in the Poynting vector (sometimes maligned as the "disapoynting" vector), which specifies the magnitude and direction of the flow of electromagnetic energy (including that which we call light). In an obituary notice about Poynting published in *Nature* in 1914, Sir Joseph Larmor opines that "his rebellion against an excessive anthropomorphism which had begun to cling around the notion of natural laws, as if they were really legal enactments to be obeyed or disobeyed by inert matter almost as if it possessed will-power and could exercise choice, some substances being praised as good radiators while others stigmatised as bad . . . Poynting's revolt against this kind of attitude to laws of nature, though doubtless more than half humorous, was in itself wholesome."

At the very least, the notion that a good absorber is a good emitter is misleading. Even if it does not mislead (my experience is that it does), the overused word *good* is meaningless unless accompanied by a criterion. Let us consider each of these points in turn.

The rate of absorption by a given object depends on its radiation environment, whereas its rate of emission depends solely on its temperature. We can imagine taking an internally heated blackbody (i.e., a "good absorber") into space. It emits infrared radiation but absorbs much less because it isn't bathed in much. Yet we can imagine exposing this same body to an intense source of infrared radiation, in which instance it would be a "good absorber" but not nearly so "good" an emitter.

What is good (or bad) depends on our aims. Suppose that we wanted to keep a sandwich warm in cooler surroundings. Because of our experiments with the flask, we would elect to wrap the sandwich in aluminum foil. Aluminum is not a "good emitter," but its "bad" emissive properties make it good for keeping the sandwich warm. To keep the sandwich cool in warmer surroundings, we could again wrap it in aluminum foil. Here are two examples in which what we might unthinkingly refer to as a bad emitter is good for our purposes: keeping a sandwich cool in warmer surroundings or warm in cooler surroundings. But we haven't exhausted the possibilities.

Our aim might be to keep an object cool in radiatively cooler surroundings. One summer in Arizona, I painted the roof of my cottage with aluminum paint so that it would reflect solar radiation, thereby giving some inexpensive cooling. In retrospect, I should have used white paint—assuming that the reflectivities for solar radiation of the two paints are about the same—because its infrared emissivity is higher than aluminum's.

To say that the surroundings on a hot Arizona day are *radiatively* cooler than a roof, which at first glance might strike you as absurd, does not imply that the temperature of the air is lower than that of the roof. My roof viewed the sky, the radiative temperature of which is that of a blackbody that emits the same amount of radiation. As a rough estimate, the radiative temperature of the clear sky is about 250°K. This will vary up or down depending on the absolute humidity and actual air temperature, by which I mean a vertically averaged temperature, not that of the air adjacent to the roof.

A final possibility is that we want to keep an object warm in radiatively warmer surroundings. An example is a transatlantic hot-air balloon. At night, the balloon must be kept warm so that it doesn't sink. Since temperature usually decreases with height and water is nearly a blackbody, the balloon's temperature can be less than that of the sea below and still be positively buoyant. To keep the air in the balloon warm, its bottom half should have a high infrared emissivity, whereas that of the top half should be low because its surroundings are radiatively cooler.

I am told that the top halves of transatlantic balloons are aluminized and their bottom halves are painted black. The reflecting top keeps the balloon from getting too cold at night and too warm during the day. (A white top would be even better during the day, but not at night.) My guess is that it is unnecessary to paint the bottom of the balloon black since it, like a glass flask or white paint, is probably already black in the infrared. The black paint is more psychologically than physically beneficial.

Emissivity and Climate Change

Many textbooks convey that emissivity is a surface property, which indeed it may be considered for opaque bodies. Both a flask and aluminum foil are sufficiently opaque that their emissivities are surface properties: slice them thinner (within limits), and their emissivities would not change.

The atmosphere is different from aluminum foil or a glass slab. Unlike the latter, the atmosphere has no palpable surface. Nevertheless, the atmosphere has an emissivity, the steady increase of which is what fuels global warming.

Another difference between the atmosphere and glass or foil is that the atmosphere's infrared reflectivity is negligible, aluminum is highly reflecting from visible to radio wavelengths, and glasses have spectral regions of fairly high infrared reflectivities. Whereas the emissivity of aluminum foil is one minus its reflectivity, that of the atmosphere is one minus its transmissivity. To say that the emissivity of the atmosphere is increasing because of increasing concentrations of carbon dioxide (as well as water vapor, methane, and other gases) is therefore the same as saying that its transmissivity is decreasing or that it is becoming more opaque.

On average, the amount of surface radiation absorbed by the atmosphere must equal that emitted by it, so the temperatures of both are linked. Since the surface is heated both by solar radiation and by atmospheric infrared radiation, the average temperature of the surface is higher than that of the atmosphere.

The solar radiation absorbed by our planet must be balanced on average by infrared radiation emitted upward by the atmosphere together with that emitted by the surface and transmitted by the atmosphere. Thus if the absorbed solar radiation is fixed, an increased atmospheric emissivity must be accompanied by temperature changes to keep the infrared flux constant. Its two components depend on atmospheric emissivity in opposite ways: emission by the atmosphere goes up with emissivity, and transmission of surface radiation by the atmosphere goes down. Since the atmospheric temperature is less than that of the surface, the temperature of the latter must increase to compensate for an increased emissivity (i.e., decreased transmissivity).

Emissivity enters the greenhouse effect by another route. A possible consequence of global warming is a greater water vapor flux into the atmosphere from the oceans, hence more cloud cover accompanied by a lower net solar radiation flux to the planet. Although more clouds would indeed reflect more solar radiation, thereby cooling the planet (a negative feedback), the emissivity of clouds is greater than that of clear air (a positive feedback). Moreover, these feedbacks depend not only on how much cloud cover is increased, but where it occurs. For simplicity, divide the troposphere (which contains almost all of the mass of the atmosphere) into two parts, lower and upper. The infrared emissivity of the lower troposphere is larger than that of the upper. Clouds added to the lower troposphere (which is already nearly black) would therefore have less effect than clouds (e.g., cirrus clouds) added to the upper troposphere.

Which of the two feedbacks—negative because of increased solar reflection by clouds or positive because of decreased infrared transmission by clouds—will dominate is uncertain. The current conventional wisdom favors net warming from increased cloud cover, but this could change after the latest results from computer models are rushed to press.

✳8

Boil and Bubble, Toil and Trouble

"The time has come," the Walrus said,
"To talk of many things:
Of shoes—and ships—and sealing wax—
Of cabbages—and kings—
And why the sea is boiling hot—
And whether pigs have wings."
LEWIS CARROLL: *Through the Looking-Glass*

*T*he time indeed has come to talk of why the sea is boiling hot, which I often am told, although not in so many words. Three students with furrowed brows once appeared at my door. They were puzzled about boiling, which was no surprise to me. One of them recited what in various forms has long vexed me: water is raised to its boiling point and *then* begins to evaporate. Can this be true? If it were, wouldn't the sea be a giant bouillabaisse from which we could ladle lobsters and crabs already cooked?

For several years I have been collecting assertions in print about the boiling point that violate common sense. Even the author of a book I admire asserts that the boiling temperature is that "at which the liquid-to-gas phase transition occurs." This is not an isolated example. I have found others that convey the notion that a liquid brought to its boiling point suddenly begins to evaporate or undergoes sharp changes in its properties.

How do we reconcile what we know from experience about evaporation and boiling, including what is commonly meant by the term boiling, with the paradoxes presented by those whose writings seem otherwise paradox-free?

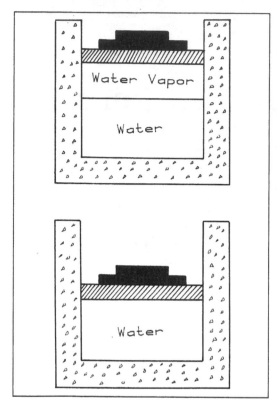

Figure 8.1
This insulated cylinder contains only water. Initially, a weighted piston rests on the surface of the water (bottom). When the water is heated sufficiently, the piston is lifted and separated from the water by a region containing water vapor (top). The temperature of the water when this occurs is called its boiling point.

The Ideal and the Real

To answer this question, I begin with the kind of thought-experiment found in textbooks of thermodynamics, a field that has never completely lost the smell of engines. Consider a cylinder containing only water and fitted with a movable weighted piston (Figure 8.1). The cylinder is surrounded by heating coils and wrapped with thick insulation, thereby allowing us to heat the water. The entire apparatus is placed in a high vacuum so that the pressure of surrounding air does not act on the piston.

Initially, the temperature of the water is such that the piston rests on its surface, hence no vapor can be above it. We turn on the heating coils and the water temperature rises. Eventually it will reach a point where the piston is lifted above the water surface and a vapor-filled space between the two is created. When this occurs, we may say without error that the water has begun to evaporate, and we may call the corresponding temperature the *boiling point*—the temperature at which the liquid and its vapor are in equilibrium. The vapor pressure is the weight of the piston divided by its cross-sectional area. If we increase this weight, the boiling point increases.

Assertions about water beginning to evaporate at its boiling point made with this ideal experiment in mind cannot be faulted on logical grounds. Nevertheless, they are unsatisfying because they seem remote from our experience of boiling, which we associate with *ebullition*, the formation of bubbles. Moreover, water in our daily lives is rarely confined in enclosed cylinders; it freely evaporates into the air above it at *all* temperatures (yes, even when the water is frozen), whereas ebullition occurs only at special temperatures (in a given environment).

Early in our education we are taught that the boiling point of water is fixed, thus providing a means for calibrating thermometers. Yet the weighted-piston definition of the boiling point implies that it can take any value because it changes with the piston weight. Given this definition, even ice can be assigned a boiling point, since the water in the cylinder can be replaced by ice. Although a boiling point for ice is logically defensible, it would surely evoke snickers in everyday conversation. It is true that the boiling point can be any temperature, but to understand why it usually is not, we must abandon ideal cylinders and pistons and enter the real world.

Boiling Means Bubbles

We know when water heating on a stove is ready for tea because of the vigorous bubbling we associate with boiling. Why is it that the water temperature must increase before bubbles form? And when they do, why doesn't the temperature increase further—or does it?

For a vapor bubble to exist in water, the pressure inside the bubble must be greater than that outside because of surface tension. This outside pressure is that of the atmosphere plus a small contribution that increases with depth into the water.

The required pressure difference across a bubble is inversely proportional to its size, which at first glance presents us with a puzzle. If bubbles grow from very small sizes, their demands must have been insatiable at the earliest stages of growth. Fortunately, nature usually provides plenty of nuclei that can serve as embryos for bubble growth. Bubbles in boiling water begin their lives not as single molecules or small aggregations of them but rather as tiny air bubbles. A bubble in water is the inverse of a cloud droplet in air. Both require nucleation to grow under conditions we have come to consider normal, even inevitable.

Bubble Growth in Water

If you heat water in a pan, you will notice small bubbles forming on its bottom and sides. These are air bubbles originating from tiny cracks and pits. When water is poured into a pan, to our eyes it is filled, yet air in invisible cracks is not displaced by the water. The air bubbles trapped in these cracks are in equilibrium with air dissolved in the surrounding water. As the temperature rises, the volume of a bubble will increase, but only by about one-third, in going from room temperature to the boiling point. Bubbles grow appreciably only as the result of diffusion of air into them.

When water is heated, its solubility decreases (Figure 8.2). The concentration of air in the water at the surface of a bubble is that appropriate to the pressure in the bubble and the new solubility. But the concentration of air some distance from this surface remains what it was before the water was heated. Thus heating the water causes a gradient of the concentration of dissolved air in the water around a bubble and as a consequence air diffuses into it.

To show how important solubility can be in determining the growth rate of bubbles, I filled a small saucepan, pitted from years of use, with tap water. Then I allowed the water to sit for about an hour. During this time the temperature of the water remained at about 20°C (68°F). The solubility of air in water at this temperature is about 19 cc/liter. No visible bubbles formed during this period. I then heated the water to about 60°C (140°F), at which temperature the solubility is slightly more than 12 cc/liter. In a short time, the bottom of the saucepan became covered with visible bubbles.

I must be careful in saying that a decrease in solubility results in an increase in the growth rate of bubbles. Without qualification, this statement may be misleading. The solubility of a gas is the amount dissolved *in equilibrium* with that above it. When

Figure 8.2
The solubility of air in water is the volume of air (at standard temperature and pressure) dissolved in a given volume of water when the air in it is in equilibrium with that above it: the rate at which air molecules enter the water is balanced by the rate at which they leave.

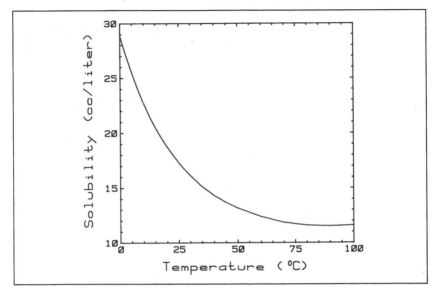

water is heated, the solubility of air in the water, but not the amount of air dissolved in it, decreases instantaneously. But it takes time for the dissolved air to again come into equilibrium. Solubility is an equilibrium property, whereas the growth of bubbles occurs because of disequilibrium. If the solubility of air in water were half, say, of its actual value, air bubbles would grow more slowly in water. But by suddenly reducing the solubility by heating, the growth rate increases because the dissolved air concentration at the surface of a bubble depends on the solubility at the new temperature, whereas that in the bulk liquid is what it was before the water was heated.

Figure 8.2 casts doubts on a statement I have seen many times in various forms. For example, I found the following in a scientific paper: "Dissolved air was eliminated . . . by using recently boiled water." The implication of this and similar statements is that by boiling water, all dissolved air can be purged from it; whereas in heating water from room temperature to the boiling point, one can only reduce dissolved air, not eliminate it entirely.

When a pan filled with water is heated, small pockets of air in cracks grow large enough by diffusion to break loose from their moorings and rise as bubbles. But they do not grow perceptibly during their ascent, at least when the temperature is low.

These small bubbles contain not only dry air (mostly oxygen and nitrogen) but water vapor as well. As the temperature of the surrounding water increases, the vapor's contribution to the total pressure in a bubble increases. When the vapor pressure is large enough, the bubble grows, and as it does the vapor pressure dominates the total internal pressure. A bubble begins its life as mostly air mixed with a little water vapor and ends it as mostly water vapor mixed with a little air.

Before boiling becomes vigorous, bubbles formed on the hot bottom and sides of the pan rise into colder water; hence, the bubbles collapse. The water temperature is not high enough to maintain a pressure in the bubble greater than that of its surroundings. But as the water is heated more, its temperature becomes uniformly high enough to allow a bubble to rise from its nucleation site to the surface without collapsing. Indeed, bubbles formed in fully developed boiling grow as they ascend, which is a clue to why the temperature of boiling water is higher than what we call the boiling point.

Can Boiling Water Rise above Its Boiling Point?

During my school years I was told that at sea level water boils at 212°F (100°C) and that once it reaches this temperature, continued heating is powerless to effect an increase. I wanted to know why, but no one could tell me. An upper limit to the temperature of boiling water just seemed to be one of the inscrutable mysteries of nature. With age, I have come to believe that unequivocal assertions about the impossibility of something, unaccompanied by any solid reason why, are likely to be false. And so it is with the supposed limit on the boiling point of water.

To better understand why, let us return to the thought-experiment. Suppose that the piston is at a fixed height in the cylinder; the water is in equilibrium with its vapor, hence at its boiling point. Suddenly the heating coil is turned on and the piston begins to rise. We are often told that once water reaches its boiling point, further heating is not accompanied by a temperature rise. Yet this cannot be true, which the experiment itself indicates. For the piston to move upward, a net force must be exerted on it: its weight per unit area must be less than the pressure exerted on it by the vapor below it. In turn, this means that the temperature of the liquid must have risen: evaporation from the water must have exceeded condensation from the vapor in order to increase the vapor pressure so that it impels the piston upward. And so it is also with boiling water. Bubbles growing in it cannot be in equilibrium, thus the water must be at a temperature higher than the boiling point.

This is hardly a recent discovery, although it seems not to have found its way into modern textbooks. In 1874, John Aitken published a delightful paper entitled "On boiling, condensing, freezing, and melting." This paper is included in his *Collected Scientific Papers*, a treasure trove for students of the atmosphere. There are no equations in any of his papers—just unadorned, clear accounts of many painstaking yet simple experiments and perceptive observations. Aitken was a pioneer in the study of nucleation. The smallest atmospheric particles, those with radii between about 0.01 and 0.1 μm, on which cloud droplets condense are often called Aitken nuclei.

When I first read Aitken's paper on boiling, I was amused by what I found on its very first page: "It is a well-known fact that in practice water boils at different temperatures under the same atmospheric pressure." This is followed by reference to a "Report on Thermometers," published in 1777, in which "it was recommended that in determining the boiling-point, the thermometer should not be placed in the boiling water, but in the steam above the water, as the water did not give the true boiling point."

Aitken recognized that all phase transitions are related, and in particular that "what we call the boiling, freezing, and melting points are the temperatures at which these changes take place when there is a *free surface* present." Thus to superheat water (i.e., raise its temperature above the boiling point), all we need to do is decrease the amount of free surface available.

To demonstrate this, I boiled water in a smooth glass flask. Without much effort, I was able to obtain a boiling temperature of 102°C (216°F), about three degrees greater than what it was supposed to be at my elevation. The bulb of the thermometer was just below the surface of the water, far from the intensely heated bottom. Moreover, the temperature of the boiling water increased with time. This is what I expected: the more air expelled from the water, the less air for nucleating bubbles.

Aitken would not have been surprised by my observations. He noted that "water from which the gases have been expelled may be heated in polished metal vessels to a temperature far above its boiling-point, and . . . when the boiling takes place under these circumstances, it does so with wonderful violence."

I don't know what the record for superheating is, and in a sense I don't want to know. Or, rather, I would prefer that others endanger their lives trying to superheat water to its utmost. Aitken writes of water being heated to 244°F (118°C), yet this "temperature at which the water exploded . . . does not by any means indicate the limit to which temperature might be raised. . . . Under more favourable conditions it might doubtless be raised to a much higher temperature."

Even when I obtained only a few degrees of superheat with my glass flask, the results were slightly unnerving. Boiling would cease momentarily, and when it resumed it would do so "with wonderful violence."

Water on the boil is heated from below and cooled from above. It cools evaporatively by vapor bubbles that rise to the surface, burst, and release their vapor. This is why it is difficult to greatly increase the temperature of boiling water in the kinds of scratched pots and pans we normally use. An increased heating rate results in greater ebullition and thus greater evaporative cooling. This feedback mechanism keeps water from heating greatly above its nominal boiling point. But it is a mechanism that works only when nucleation sites are in abundance.

Why Add Salt to Your Spaghetti Water?

Up to this point I have tacitly assumed that the boiling water is pure. What if it is not?

The saturation vapor pressure of water decreases when anything is dissolved in it. A solute decreases the number density of water molecules, hence their rate of evaporation. Saturation vapor pressure is a measure of evaporation; if the latter decreases so does the former. The nature of the solute is largely irrelevant, especially at low concentrations where lowering of the vapor pressure is determined solely by the amount of solute dissolved, not by its chemical composition.

The equilibrium vapor pressure of a bubble surrounded by salty water is less than that of a bubble surrounded by pure water, both at the same temperature. Thus for a bubble not to be crushed in salty water, the water's temperature must be raised above what would have been necessary had the water been pure.

When I ask my classes why salt is added to water for cooking pasta, I almost invariably get two answers. I am told that adding salt decreases the time it takes water to boil. Yet if salt increases the boiling point, water takes longer to boil. I am also told that salt increases the boiling point, which is correct but largely immaterial. To determine how much the boiling temperature rises with the addition of salt, I ladled great heaps of it into boiling water. Its temperature increased to only 108°C (226°F). Anything cooked in this brine solution would be unpalatable. So I conclude that salt is solely for flavor in the quantities added to cooking water, which is what Jearl Walker concluded in the December 1982 issue of *Scientific American*. This chapter complements his article on boiling, which I recommend. It is especially good on the formation of vapor bubbles and their dynamics.

Perhaps one reason my students think that adding salt to water decreases the time to bring it to a boil is that when salt is added to boiling water, it may momentarily boil more vigorously. This is because the salt grains, before they dissolve, provide many nucleation sites.

Ebullition Altimetry

The boiling point of water or any liquid depends on the local atmospheric pressure. If vapor bubbles can survive in water only if it is hot enough to provide a vapor pressure in excess of the surrounding pressure, the lower this pressure is, the lower the boiling point will be. Determining the boiling point of water is thus a means for measuring elevations because atmospheric pressure decreases with height.

Figure 8.3 shows the boiling point of water as a function of elevation in a standard atmosphere—which does not exist (it is

Figure 8.3
The boiling point of water depends on the pressure of its surroundings. This pressure decreases with elevation, hence so does the boiling point. Under standard atmospheric conditions, the boiling point decreases linearly with elevation at a rate of about 3.33°C per kilometer.

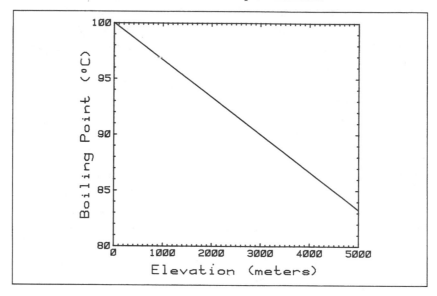

not kept in a glass case at the National Bureau of Standards). The standard atmosphere is a hypothetical atmosphere upon which a group of scientists once agreed but would probably not agree again. The standard atmospheric sea-level pressure is 1013.25 mb; the temperature is 15°C (59°F) and decreases at a rate of 6.5°C per kilometer.

Note that the decrease of the boiling point with elevation is nearly linear. This is a happy accident. Saturation vapor pressure increases approximately exponentially with temperature (Figure 8.4) and air pressure decreases approximately exponentially with elevation (Figure 8.5) over the range of interest. As a consequence, the lapse rate of the boiling point (i.e., its rate of decrease with elevation) is a nearly constant value of about 3.33°C per kilometer (1.83°F per 1000 feet). Although this lapse rate is for a standard atmosphere, it is nevertheless a good rule of thumb.

I live at an elevation of about 414 meters below Mt. Nittany, the elevation of which is about 631 meters. The boiling point of water at my home should be (under standard conditions) 98.6°C (209°F); that at the top of Mt. Nittany should be 97.9°C (208°F).

Figure 8.4
The saturation vapor pressure of water increases almost exponentially with temperature.

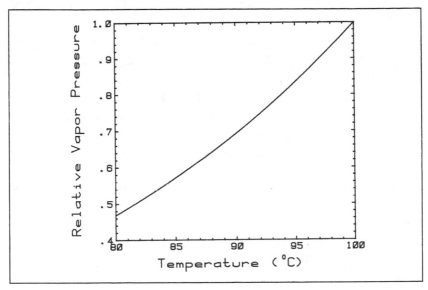

Figure 8.5
Atmospheric pressure decreases almost exponentially with elevation.

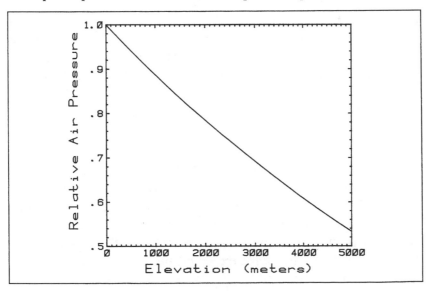

This decrease is large enough to be measured with an inexpensive thermometer, so I decided to give it a try. But first I had to devise a scheme for measuring the boiling point with as little error as possible.

Simplistic discussions of the boiling point convey the impression that one merely need plunge a thermometer into a boiling liquid and the thermometer will infallibly register the boiling point. I stated previously that I measured a temperature of 102°C in a flask of boiling water. This was the maximum temperature. Depending on the depth of the bulb below the water surface, the boiling temperature ranged from 99 to 102°C, and only small changes in this depth caused the reading to vary by a degree or more. This variation could swamp my expected 0.7°C temperature change if I were not careful.

For measuring the boiling point, I selected an ancient saucepan, scarred from years of use. I mounted my thermometer on a stand and positioned it so that the bulb was as nearly horizontal as possible just above the surface of the water. Much of the bulb was exposed uniformly to hot vapor, which condensed on it. So I was really measuring the *condensation* temperature, not the higher boiling temperature (i.e., the temperature of the boiling liquid). I covered the pan and thermometer with a sheet of aluminum foil to suppress convection and (net) radiation from the bulb; polished aluminum reflects the infrared radiation emitted by the bulb (see Chapter 7). With this setup, I was reasonably confident that I could measure the expected temperature change. The smallest graduation on my thermometer was 1°C, but by using a magnifying glass I could estimate temperatures to within 0.2°C.

One February morning I went out onto our porch and boiled some water in my covered saucepan heated with a small backpacker's stove. The condensation temperature was 99°C, somewhat higher than expected. But the clear blue sky indicated that the pressure might be higher than average. And indeed it was, about 2.5 percent. Such a pressure increase corresponds to an increase in the boiling temperature of about 0.6°C. Thus the corrected boiling temperature was 99.2°C, close enough to what I measured. I couldn't have expected better given the precision of my thermometer.

Then I stowed all my gear in a rucksack and trudged up Mt. Nittany, taking my dog with me after making him promise not to upset my stove with his wagging tail. At the top, I measured a condensation temperature of 98.5°C, about what I expected. Using my somewhat crude apparatus, I can measure elevations to within about 60 meters. I wouldn't want to fly in an airplane equipped with an altimeter no better than this, but for some purposes it would be adequate.

If you take a backpacking trip in the mountains, I suggest that you bring with you a thermometer capable of measuring temperatures near 100°C. By carefully determining the condensation temperature along your route, you should be able to estimate your elevation. If you have a topographic map, you might even be able to find your way if you get lost.

*9

An Essay on Dew

*When I come to die I shall not feel
that I have lived in vain. I have
seen the earth turn red at evening,
the dew sparkling in the morning, and
the snow shining under a frosty sun.*
BERTRAND RUSSELL

*I*t was a bit daunting to contemplate writing about dew after reading the opening sentence of John Aitken's 1885 paper "On Dew": "The immense amount that has been written on the subject of dew renders it extremely difficult for one to state anything regarding it which has not been previously expressed in some form." Nevertheless, I was heartened that Aitken's own admission did not deter him from adding more than 50 pages to the "immense amount" already published.

Aitken did have several predecessors, even excluding those who wrote "from the purely literary point of view." You can read about them in W. E. Knowles Middleton's excellent book *A History of the Theories of Rain*, Chapter 9 of which summarizes the history of theories of dew and frost, beginning with Aristotle and ending with Aitken. I regret that I must leave the distant past in Middleton's capable hands and proceed to that more recent.

A Controlled Experiment Is Preferable to a Shouting Match

As so often has happened, the experiment described in this chapter arose out of my teaching. One fall I would casually notice dew on my truck each morning at the start of my hike over Mt. Nittany to a nearby town. Later in the day, in class, I would recount where dew had formed, how much, and sky conditions the previous night. Sometimes my observations would accord

97

with ones that had been made by my students, other times not. This caused me to be more observant. For example, on some mornings when no dew was apparent at first glance, wet fingers would tell me that it is easy to miss a fine coat of dew on a white truck. I soon realized that I would never resolve the seeming contradictions that arose in classroom discussions without some experiments. Thus I progressed from casual to careful observations and finally to controlled ones.

One of my classroom assertions was that I expected dew (or frost) to form more readily on thermal insulators such as glass and wood than on conductors such as metals, all else being equal. The qualifier is crucial because several factors determine whether dew will form on an object: its orientation, its composition, and its surface properties. Added to these attributes of the object are those of its surroundings. So I did some experiments in which as many of these factors as possible were kept constant.

I would have liked to do my experiments with large, thick slabs, but I had to settle for plates about eight millimeters thick. One plate was aluminum and two were wood; I did my best to make them otherwise identical in all respects—thickness, surface condition, and surface area. After the coat of black paint I gave them had dried, I put them on a large flat rock open to the sky. Then early every morning I would examine them for dew or frost. Often none had formed, or my experiment was ruined by rain or snow. But on clear, calm nights when dew or frost did form, it usually was thickest on the wooden plates (see Figure 9.1). One morning, however, I was puzzled to find more dew on the aluminum plate than on the wooden plates. My first thought was that water had seeped into the more porous wood, so I gave all the plates a coat of clear varnish to seal them. Some time later I mentioned this slightly puzzling observation and my interpretation of it in class. Cliff Dungey immediately responded with, "Are you sure that it didn't get warmer overnight?" His question jarred me out of my complacency. That January morning had been warmer than the previous evening, which was the real reason for the greater amount of dew on the aluminum plate. Before discussing this and other results, we need to lay some foundations.

Figure 9.1
The center plate is aluminum, those flanking it are wood. All of them are about the same size and were painted the same. More frost forms on the wooden plates because their much lower thermal conductivity allows their surface temperatures to drop lower than that of the aluminum plate. The aluminum plate's edges are lined with frost because the surface area there relative to the volume of surrounding metal is greater than that away from the edges (see Figures 9.7 and 9.8).

Dew Formation

The generally accepted definition of the dew point is that given in Chapter 1: the temperature to which moist air must be cooled, at constant pressure, for saturation to occur. Saturation here is that relative to liquid water. If that relative to ice is meant, the corresponding temperature is called the frost point. For brevity, I shall henceforth discuss mostly dew, although my comments apply equally well to frost.

For dew to form on a surface, its temperature must drop below that of the dew point of the air to which it is exposed. This statement is not incorrect, but it is a bit unsatisfying. It is more a definition of the dew point than an explanation of why water accumulates on cold surfaces. One sometimes finds the statement that dew forms when air is cooled below its dew point, which is misleading: it is the surface that must be cooled.

The way to understand the formation of dew (and cloud droplets) is to recognize that the rate of condensation is proportional to the number density of the water vapor component of air. Water molecules continually bombard surfaces exposed to air, some of them stick, and a thin film of water accumulates. It may be momentary, however, because molecules in this film are in a state of continual and rapid agitation. Some of them therefore leave the film. The rate at which they leave depends on the film's temperature, essentially the same temperature as that of the surface on which the film formed.

Two processes are going on continually: condensation—the sticking of water molecules randomly bombarding a surface—and evaporation—the escape of molecules from a deposited water film. The latter depends on the state of the surface, the former on that of its surroundings. If the rate of condensation exceeds that of evaporation, water accumulates, resulting in *net* condensation. The lower the temperature of the surface, the lower the rate of evaporation of water from it. At some temperature the evaporation rate must drop below the condensation rate with the result that liquid water accretes on the surface.

All this is easy enough to state and, I hope, to understand. Yet many questions remain unanswered. Why does dew form some nights but not others? And why does it form mostly at night? What is it about clear, calm nights that favors dew? Why does it form on some surfaces but not on neighboring ones?

Nocturnal Radiational Cooling

All objects at temperatures above absolute zero emit radiant energy of all wavelengths, an inexorable and incessant process occurring day and night for eternity. Such radiation is often called *heat radiation*, a term that causes me to cringe. And I am not alone, as I discovered in *Light, Colour, and Vision*, the author of which, Yves Le Grand, does not conceal his contempt for "the meaningless term *radiant heat*; to say that the sun . . . radiates heat is naive." He says further that "the expression *heat rays . . .* is doubly misleading. . . . Radiations of equal energy, if equally absorbed, give rise to the same amount of heat[ing] no matter what their wavelength." Here is a man after my own heart, one who doesn't shrink from using words such as *meaningless* and *naive*.

100

I cringe even more when I encounter statements conveying the notion that matter is equipped with sensors and switches to turn it on at dusk. Objects do not begin radiating at night. Indeed, since they are usually warmer during the day, they radiate even more then than at night. And although they are no longer warmed by the sun at night, they still are warmed by radiation from their surroundings, including the sky.

On clear nights, however, an object may emit more radiation than it receives from the sky, thereby cooling (in the absence of other mechanisms for energy transfer, a point to which I shall return). This was recognized as long ago as 1814 by William Charles Wells, who stated in his "Essay on Dew" that "[w]hatever diminishes the view of the sky, as seen from the exposed body, occasions the quantity of dew which is formed upon it to be less than would have occurred if the exposure to the sky had been complete."

The key words here are *view of the sky*. Imagine a flat plate with a spider on it. What the spider sees depends on the orientation of the plate. We may divide her field of view into two parts, that above her horizon and that below. When the plate is horizontal and she is on its upper surface, her field of view lies entirely above the horizon. When the plate is vertical, half her field of view lies above the horizon, half below. When the plate is horizontal and she is crawling on its underside, her field of view lies entirely below the horizon.

The region below the horizon is usually dominated by the ground, buildings, and vegetation, whereas that above the horizon may be dominated by the sky, especially in open areas. Often, the ground is warmer than the sky. I must be careful to distinguish the radiative temperature of the sky from the air temperature: the two are not the same. Imagine the sky to be replaced by a blackbody, the temperature of which is such that it emits the same amount of radiation as the sky. This temperature is its radiative temperature.

A good rule of thumb is that the clear sky has a radiative temperature of about 250°K. If we confine our attention to nights with air temperatures near or above freezing, the temperature of the ground will be greater than 250°K. Under these conditions, more radiation is emitted by the ground than by the sky. So if

Figure 9.2
The field of view of every point on this post determines the pattern of frost on it. Radiational warming by the ground was sufficient to keep the sides above the frost point.

our spider wants to keep as warm as possible, she'll stay on the underside of a horizontal plate.

The consequences of unequal radiative temperatures of ground and sky—and different views of both, depending on orientation—to the formation of dew and frost are evident in Figure 9.2, which shows a dark post that has acquired overnight a nonuniform coating of frost. The top of the post has a coating, whereas its sides are frost free—evidence of the different radiative environments of the top and sides. Radiative cooling was insufficient to lower the temperature of the sides below the frost point. It is scenes such as this post that undoubtedly are responsible for engendering the misconception that dew and frost fall from the sky just like rain and snow.

Nocturnal Cooling of Metals

Most natural objects, such as soils, snow, and trees, are nearly blackbodies. The fact that this may not be true, however, for some unnatural objects such as metals is significant when one considers the formation of dew. The earliest investigators observed that polished metals did not collect nearly as much dew as did materials such as glass, both in the same environment. Not surprisingly, this led them to believe that metals repel dew. In a certain sense they were right, depending on what is meant by *repel*. Dew is indeed less likely to form on a polished metal surface, but not because it wards off water molecules impinging on it.

Polished metals such as aluminum and silver are highly reflecting at visible wavelengths. They also are highly reflecting for radiation of wavelengths longer than those of visible light, in particular the infrared radiation emitted by objects at normal terrestrial temperatures.

I showed in Chapter 7 that the infrared emissivity of aluminum foil is less than that of glass or of painted surfaces. Hence aluminum foil—and indeed, most polished metal surfaces—both emits and absorbs at a lower rate than insulators at the same temperature and in the same environment. Cooling occurs if emission is greater than absorption, the difference between which is less for polished metals. This explains, at least in part, observations of dew formation mentioned at the beginning of this section. You might be tempted to conclude that this is why less dew formed on the aluminum plate of my experiment than on the wooden ones. Recall, however, that I painted the wood and the aluminum so that they appeared identical. Of course, merely because a painted surface appears black to our eyes does not necessarily mean that it is black in the infrared. But the experiments in Chapter 7 provide evidence that the aluminum does not shine through the black paint at infrared wavelengths. Thus we have to turn elsewhere for an explanation of what I observed.

From a fundamental point of view, the reasons why polished metals are comparatively feeble emitters and why less frost formed on the aluminum plate are the same. A metal is, so to speak, a box of mobile or free electrons, which transport both electric charge and energy much more readily than bound electrons in, say, glass or wood. The high electrical conductivities of metals are associated with their high reflectivities. And high *electrical* conductivity is associated with high *thermal* conductivity.

Energy transfer because of temperature differences in solids (indeed in all sensibly stationary matter) is described by the Fourier heat conduction law, which states that the time rate of energy transfer is proportional to the rate at which temperature changes with distance (the temperature gradient). The proportionality factor, called the thermal conductivity, varies greatly for different materials. For example, it is about 2,000 times greater for aluminum than for wood.

Consider two slabs, infinite in thickness and in lateral extent, one wood, the other aluminum. They both are at the same uniform temperature and are being warmed radiatively at the same rate they are being cooled. Suddenly the balance is tipped in favor of cooling and the surface temperature of each begins to decrease. This results in a temperature gradient near their surfaces. Fourier's law implies an upward energy flux, which, for equal temperature gradients, is much greater in aluminum than in wood. This upward flux of energy warms their surfaces, which are cooled by radiant emission but warmed by conduction from below. For the same surface conditions, radiative cooling is the same, but conductive warming occurs at a much greater rate in aluminum than in wood. As a consequence, the surface temperature of the wood drops, whereas that of the aluminum remains nearly constant (see Figure 9.3). The high thermal conductivity of aluminum ensures that any temperature gradients in it are rapidly suppressed, whereas large temperature gradients can develop in insulators such as wood (see Figure 9.4).

The calculations shown in Figures 9.3 and 9.4 were done for infinite slabs, although the plates of my experiment were finite. There certainly are differences between finite and infinite slabs, but this does not invalidate my physical arguments. Because of its high thermal conductivity, aluminum cools in such a way that its temperature is nearly uniform, whereas the much lower thermal conductivity of wood allows its surface temperature to drop well below that of its interior.

One corollary of my experimental results and their interpretation is that previous statements about dew formation on polished metal surfaces seem to have missed the essence of what happens. If dew does not form on such surfaces, is this attributable to low emissive power or to high conductivity? My metal

Figure 9.3
The surface of an object will cool to the temperature of its surroundings
at a rate depending on the thermal conductivity of the object. The relative
temperature difference is the difference between the temperature of the
surface and that of its assumed colder surroundings relative to the initial
value of this difference. Note that the rate of cooling decreases after a few
hours. The curve shown is for wood. The relative temperature difference
for aluminum (not shown) is imperceptibly different from unity over the
time interval shown.

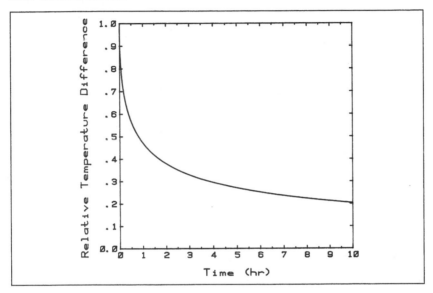

plate was painted so that its emissive power was the same as that
of the wooden ones, yet dew still formed less readily on the metal
one, which points to the second of these two reasons as the more
important.

Now we are better able to understand Cliff Dungey's rhetor-
ical question about my puzzling observation. For the same reason
that dew forms more readily on wood upon cooling, it also dis-
sipates more readily upon warming. The morning on which I
observed the absence of dew on the wooden plates was warmer
than the previous evening. What probably happened is that dew
formed on all the plates, then warm air was advected into my
outdoor laboratory and the surface temperature of the wooden
plates increased more rapidly. Hence the dew on the wooden
plates evaporated faster than did the dew on the aluminum plate.

Figure 9.4
After two hours of cooling, a large temperature gradient near the surface has been established in wood (solid curve), whereas in aluminum (dashed curve) the temperature is still uniform.

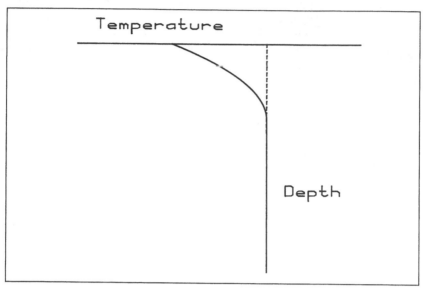

In the Calm of the Night

Early in this chapter I mentioned, without further comment, that dew forms on "clear, calm nights." The first qualifier is related to radiative cooling: clouds generally have a higher radiative temperature than the clear sky has. Why do calm nights favor dew formation?

Objects in the night air exchange energy with their surroundings not only by radiation and conduction but by convection as well. They are in contact with air, which usually is warmer than their surfaces. Hence there is a convective flux of energy to surfaces, a radiative flux of energy from them, and a conductive flux to or from them.

All modes of energy exchange occur simultaneously, but one or the other may dominate. The higher the wind speed, the more convection dominates, and the closer the surface temperature will be to the air temperature. For dew to form, the surface temperature must be below the dew point, which in turn is less than the air temperature. So the greater the wind speed, the less likely the surface temperature is to drop below the dew point.

(An addition to my collection of traps people fall into because of arrogance was acquired during a visit to a European university. My host told me an amusing story. He has been developing coatings for applications in radiative heating and cooling. At a presentation of his work, he was lambasted by a red-faced, fuming professor of theoretical physics who stated categorically that it is "thermodynamically impossible" for a body to cool below air temperature. It took the efforts of a mere professor of agriculture to convince this unobservant soul, who had mastered thermodynamics but never had bothered to observe frost form, that objects frequently cool by radiant emission to temperatures lower than that of the surrounding air.)

There is another reason that dew doesn't form on windy nights. The higher the wind speed, the more that warmer air will be mixed with colder air below it, thereby keeping the surface warmer than it would otherwise be. And if this warm air is drier than the surface air, dew is further suppressed. What favors the formation of dew is a low surface temperture and a high dew point: wind can work against both.

The consequences of a high dew point can be observed by inverting a bucket on a lawn overnight. Although dew may not form on the bucket's outside surface, the surface inside may be dripping wet. This seems to contradict my assertion that a surface with a view of the ground usually undergoes less cooling than one with a view of the sky. One must remember, however, that in an enclosed space (or even merely in stagnant air) the vapor density increases (assuming a plentiful supply of water) until it is close to the saturation value. Thus the dew point in the moist air under the bucket is greater than that in the drier air surrounding it.

I recommend the following variation on my experiment. At sundown on a day that promises a clear, calm night, set out identical plates—one metal, one wood or glass—on a hard surface. At dawn examine both the top and bottom sides of these plates. What you see may at first glance be a puzzle, but if you understood the preceding arguments, you should be able to solve it.

The Fruits of Teaching

Among the captive audience of students who have had to listen to my babblings about plates was at least one receptive listener.

Figure 9.5
The pattern of frost on this car is a consequence of the different fields of view of each part of it as well as the different materials of which it is composed. Photograph by Tracy Cox.

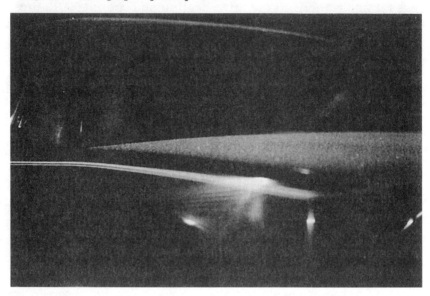

Tracy Cox, coming home from a nocturnal prowl, happened upon a frost-covered car that sent him scrambling for his camera, which he propped on a garbage can to take the time-exposure shown in Figure 9.5.

His photograph illustrates several points made in this chapter. For example, the rear quarter panel of the car is free of frost, whereas the trunk lid is thick with it. And note the sharp boundary between the frosty lid and the strip adjacent to it. This boundary is an air gap that thermally insulates the one region from the other.

Careful scrutiny reveals a gradual increase in the density of frost proceeding upward from the quarter panel along the rear pillar to the roof; the last, like the trunk lid, is coated with frost.

The rear window is clear, which is perhaps puzzling since it is a much better insulator than the surrounding metal. But this window sees more of the ground and adjacent buildings than the trunk lid does. Moreover, the window is in better thermal contact with the interior of the car than is the metal paneling, which is covered on the inside with insulation.

Tracy's photograph also illustrates that what for some people is hardly worth a second glance is for those observant few a fascinating scene, rich in detail and in scientific puzzles.

Frost at Edges

Note in Figure 9.1 that the edges of the aluminum plate are lined with frost. I could attribute this to an "edge effect," but this would be a swindle: an edge effect is an effect that occurs at an edge—true, but not very satisfying.

To explain why frost forms more readily at edges, consider a ball, small enough to be assigned a single temperature, cooling in air. The energy of this ball is proportional to its volume (or mass) and its temperature. Since the volume of the ball does not change appreciably as it cools, the decrease of its energy is proportional to that of its temperature. The greater the ball's volume, the smaller its temperature change as a result of a given decrease of energy.

The ball cools because it exchanges energy with its surroundings at a lower temperature. This energy exchange is proportional to the difference between the ball's temperature and that of its surroundings as well as to the surface area of the ball.

In any time interval, the decrease in the energy of the ball because of interaction with its surroundings is proportional to the ball's surface area. But for a given energy decrease, the corresponding temperature decrease of the ball is smaller the larger its volume. Thus the temperature decrease in any time interval is directly proportional to the ball's area and inversely proportional to its volume. The rate of cooling of the ball therefore proceeds faster the greater its surface-to-volume ratio.

Examples of this are abundant. When my wife and I used to drive long distances during winter, we would carry with us an assortment of thermos bottles, some large, some small, filled with coffee. We would always take coffee from the smallest bottles first because they cooled the most rapidly.

In Pennsylvania, signs warning "Bridge Freezes Before Road" (Figure 9.6) stand at the approaches to many bridges. The surface-to-volume ratio of a bridge is much greater than that of the land masses it spans.

Our fingers and toes, ears and noses, freeze before our thighs do (one must be careful discussing cooling of the human body

Figure 9.6
Signs such as this one at the approach to a bridge in Pennsylvania testify to the consequences of high surface-to-volume ratios.

Figure 9.7
The surface area of a volume at *B* is larger than that of an equal volume at *A*.

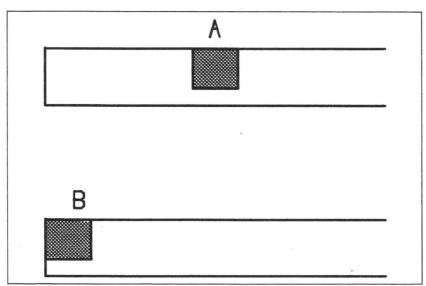

Figure 9.8
After a night of light frost, these tiny leaves displayed delicate fringes of frost in the early morning light.

since it is not a passive object like a bridge or a thermos bottle filled with coffee). Frostbitten fingers are common, but I have never heard of anyone with frostbitten thighs or buttocks.

Now let's consider the plates in my experiment with dew and frost formation. Figure 9.7 shows a plate in cross section. The exposed surface area of a given volume at *A* is less than that of the same volume at *B*; hence *B* cools more rapidly.

The consequences of high surface-to-volume ratios to the formation of frost can be observed readily, especially if you are willing to get down on your hands and knees and peer at the leaves on small plants early in the morning after a light frost. Figure 9.8 shows the kind of scene that awaits you. Tiny leaves on my lawn were fringed with frost, which soon disappeared after the sun rose above the surrounding ridge tops.

111

✳10

Mad Dogs and Englishmen Go Out in the Midday Sun

In Bangkok at twelve o'clock
They foam at the mouth and run
But mad dogs and Englishmen
Go out in the midday sun
NOEL COWARD

"The idea of a sun millions of miles in diameter and 91,000,000 miles away is silly. The sun is only 32 miles across and not more than 3,000 miles from the earth. It stands to reason it must be so. God made the sun to light the earth, and therefore must have placed it close to the task it was designed to do. What would you think of a man who built a house in Zion [Illinois] and put the lamp to light it in Kenosha, Wisconsin?"

This marvelous piece of nonsense, quoted by Martin Gardner in his superb book *Fads and Fallacies in the Name of Science*, is the product of the addled mind of Wilbur Glen Voliva, a crank who held many eccentric views, not the least of which was that the earth is flat. But the rantings of madmen often contain a germ of truth; those of Voliva are no exception. Before we probe further, however, it is well to discuss briefly how the size of the sun and its distance from the earth are related to our habitation of this planet.

The Solar Irradiance

That solar radiation is essential to the running of the Weather Machine is incontrovertible: were there no sun, weather maps

113

would be quite uninteresting (and irrelevant), for solar radiation is the prime mover of the atmosphere. It is indeed fortunate that we have a sun. Even more fortunate, the amount of radiant energy it emits and its distance from the earth are just about right. If either of these were much different, life on this planet would be a grim affair.

On a hike along the Welsh coast many years ago I was asked by Bowen George, a Welsh steelworker, whose lack of formal education did not inhibit his inquisitiveness, this question: "How much energy from the sun does Jupiter receive compared with what we receive?" To answer this, we need to know the *solar irradiance* for Jupiter compared with that for the earth.

The solar irradiance is often called the *solar constant*. But this term is embarrassing to those engaged in measuring its variability. After all, a constant is a constant is a constant, and logic forbids it to vary, although its value may not be known precisely.

Imagine being high above the earth where there are fewer molecules and particles than near the surface. And imagine pointing a one-meter–square sheet of paper toward the sun. The amount of solar radiation impinging on this paper in one second is defined as the solar irradiance. At the top of Jupiter's atmosphere this number is different. How much different, and why, is related to Bowen's question, which I address next.

Pincushion Analogy

Consider a small pincushion in which a large number of long pins are embedded and about which a sphere is drawn. Regardless of the sphere's radius, the number of pins that pierce it is fixed, but the number that pierce a fixed area (one square meter, say) is not. As the radius of the sphere is increased, the same number of pins pierce an ever larger area, so the number of pins allocated to each bit of it (of fixed size) must decrease. For a sphere of given radius, the number of pins piercing a unit area is the total number of pins divided by the area of the sphere. This area increases as the square of the sphere's radius.

The total number of pins bristling from the pincushion is analogous to the radiant power emitted by the sun in all directions, and the number of pins piercing a unit area of a sphere centered on the pincushion is analogous to the solar irradiance.

Thus the solar irradiance decreases as the square of the distance from the sun. A collector of fixed size pointed toward the sun intercepts less of its energy the farther away it is—would anyone have guessed otherwise? But we have gone somewhat beyond common sense to arrive at a quantitative relation for the dependence of the solar irradiance for a planet on its distance from the sun: it decreases inversely as the square of this distance. Jupiter is more than five times farther from the sun than is the earth; consequently, the solar irradiance for Jupiter is less than four percent of what it is for the earth. But Jupiter intercepts more solar energy, about five times as much, because its radius is about 11 times that of the earth.

Note that if the earth is to be habitable, the inverse-square relation places rather strict limits on its distance from the sun: halving it would result in an intolerably hot planet; doubling it, an intolerably cold planet. We are fortunate that the solar irradiance for the earth is what it is, neither much more nor much less. But what is it?

Estimating the Solar Irradiance

Voliva's eccentric notions about the sun are correct in at least one respect: the sun *could* be "only 32 miles across and not more than 3,000 miles from earth." Such a sun would appear the same and would give the same solar irradiance as a sun much larger but proportionately farther away. Stated another way, to build a warming fire you can either make it big and stand back or small and stand close. And herein lies the means for estimating the solar irradiance without accompanying mad dogs and Englishmen out in the midday sun. This was passed on to me by a colleague at Penn State, Alistair Fraser, who in turn got it from the famous Dutch astronomer Hendrick van de Hulst.

To estimate the solar irradiance without going outside, all that you need is a light bulb, a ruler, and a few willing volunteers, preferably ones who have made a recent trip to the beach. Ask one of them to imagine how warm the rays of a bright sun feel on the bare skin of his hand. Then bring the light bulb slowly toward it until he judges heating by the bulb to be about equal to that by the sun on the imagined warm day. Note the distance from the center of the bulb to his hand. From this the solar irradiance can be estimated as the light bulb wattage divided by the area of a sphere with a radius of the hand-bulb distance.

This can be repeated with different volunteers and an average taken. You can also explore the consequences of using either the palms or backsides of hands or skin on other parts of the body—within the limits imposed by good taste and the law.

The strict validity of the inverse-square relation requires that the source (bulb or sun) be small compared with the distance from it. This condition is certainly not satisfied by the bulb, which must be close to the skin before it warms as much as sunshine. Moreover, the source must be spherical, which is satisfied only approximately by the bulb. And the solar irradiance is defined as the amount of energy received at the *top* of the atmosphere, whereas you are relying on the judgment, highly subjective at that, of how sunshine feels at the *bottom* of the atmosphere. Despite these drawbacks, this simple method for estimating the solar irradiance usually gives values remarkably close—within about 10–20 percent—to the accepted value.

What is the accepted value? At present the best estimate is about 1370 watts per square meter. Like the price of gold and the hemline of skirts, it fluctuates, but fortunately for us, not so wildly.

Season to Taste

Many years ago I heard a radio program in which a reporter was talking to young schoolchildren, perhaps second-grade students, who had just learned about the seasons. Although they clearly explained seasonal changes to him, he responded with, "Now let me see if I've got this right. It's colder in winter because then we're farther from the sun."

"No! No!" the students shouted, their frustation almost palpable. "You've got it all wrong." And then they tried to set him straight, but to little avail. He persisted in his adherence to the solar distance theory of the seasons. To this day, I do not know if he was remarkably stupid or merely probing how well the students had learned their lesson—or perhaps just having a bit of fun with them. In any event, I have never forgotten those children. I was impressed by how firmly they held their ground despite the efforts of an adult to dislodge them.

Perhaps because of this experience, I had assumed that everyone, with the possible exception of one thick-headed reporter, understands the origins of seasonal variations in temperature. But slowly I have come to the conclusion that this is not so. From time to time I have been shaken out of my complacency by the misconceptions harbored even by meteorology students, who, one would think, ought to understand why temperatures in mid-latitudes are generally higher in summer than in winter.

It is not enough to inveigh against misconceptions; to root them out one has to get to their origins. Erroneous notions about seasonal variations may originate from the inherent difficulty of making accurate scale drawings of the earth and sun. Simple diagrams often are helpful in efficiently conveying ideas that otherwise would require many carefully crafted sentences, but diagrams have their dark side as well; they are not all goodness and light. For example, a diagram of the earth's orbit drawn accurately to scale would at first glance be almost indistinguishable from a blank sheet. A dot would represent the sun because its diameter is about one-hundredth its distance from the earth. Without the aid of a microscope, the earth would be imperceptible because the sun is about 200 times larger.

More to the point here is the shape of the earth's orbit. We are told that it is elliptical, a discovery that immortalized Keppler. But we won't believe this from a drawing unless it is distorted greatly. Figure 10.1 shows what might be called an anatomically correct drawing of the earth's orbit, although the relative sizes of the earth and sun are exaggerated. This orbit is elliptical, which you can verify for yourself with a straightedge if you have a steady hand and a good eye. But the deviation from a perfect circle is so small that it results in an orbital variation in the solar irradiance of only a few percent. This is not enough to account for the seasonal swings in mid-latitude temperatures. Moreover, the earth is closer to the sun in the northern hemisphere winter than in summer, although this is no cause for feeling especially favored over residents of the southern hemisphere. Climatic differences between conjugate points in the two hemispheres have more to do with the comparative paucity of land masses in the southern hemisphere than with small variations in the solar irradiance.

Figure 10.1

When drawn accurately to scale, the ellipticity of the earth's orbit is difficult to discern without careful inspection.

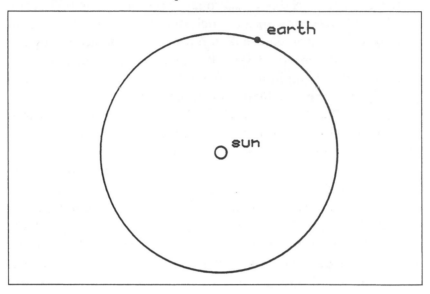

Figure 10.2

The angle between direct sunlight and the earth varies over the globe. This is the major contributor to latitudinal variations in temperature.

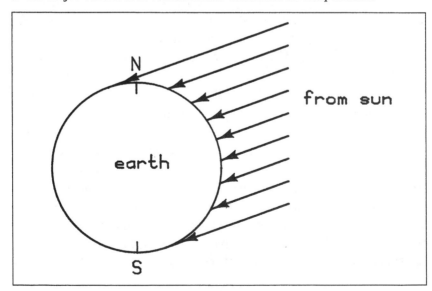

Figure 10.2 embodies the real reason behind seasonal climatic variations, although I purposely drew it so that it can be misleading. I have discovered that some of my students have been misled by figures like this one, which show that the distance from sun to earth depends on latitude. Can this be the cause of latitudinal climatic variations? The lengths of the arrows in Figure 10.2 representing rays from the sun differ by a factor of almost two. But this is only because I cannot show the earth and sun together accurately to scale. Although it is indeed true that the sun-earth distance depends on latitude, the maximum variation of this distance over the globe is less than one-hundredth of a percent, of no climatic consequence.

In giving the definition of solar irradiance, I was careful to require that the detector (a sheet of paper one meter square) be pointed toward the sun. If it is not, less solar radiation will be received by the detector, the limiting case being when it is tilted 90 degrees to the sun, when it will receive none. This can be demonstrated easily by tilting a flashlight while shining it at a wall. The amount of radiant energy from the flashlight is constant, but this energy is distributed over a larger area as the flashlight is tilted away from the perpendicular to the wall.

In Figure 10.2, rays from the sun are more oblique to the earth closer to the poles. Patches of fixed size on the earth receive less solar energy poleward than near the equator. This goes a long way toward explaining why temperatures generally decrease with increasing latitude. But this trend is not invariable: winters in Montana, for example, are brutally colder than those in Iceland, more than 15 degrees closer to the pole. Global temperature variations are also influenced by the presence of mountain ranges, the proximity to the sea, and the meandering of ocean currents. Nevertheless, with suitable caveats, it is indeed true that temperatures decrease poleward from the equator.

Yet if it were not for the approximately 23-degree tilt of the earth's axis relative to the plane of its orbit, we would experience latitudinal variations in climate but not seasonal ones. We would enjoy a climate much like that of the tropics, where seasonal variations in temperature are small (seasons in the tropics are not characterized as hot or cold but rather as wet or dry). Figure 10.3 shows the angle made by sunshine with the earth at 45°N latitude in summer and in winter. At the summer solstice (about

Figure 10.3
At the same latitude (45°N in this diagram), the sun is more nearly overhead at solar noon in June than in December because the earth's rotation axis is tilted with respect to the plane of its orbit around the sun.

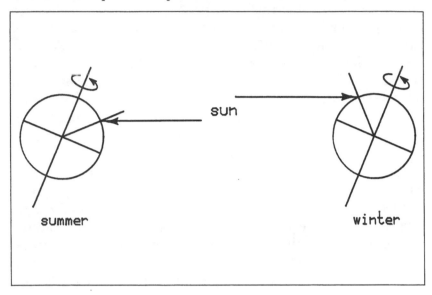

21 June), the rays of the sun at solar noon (when the sun is highest in the sky) are directed about 22 degrees from the zenith, whereas at the winter solstice (about 22 December), the rays are directed about 68 degrees from the zenith. As a consequence, a given patch of the earth receives considerably more solar radiation at the summer than at the winter solstice.

One wonders how Wilbur Glen Voliva would have explained both latitudinal and seasonal variations in temperatures over a flat earth.

*11

Temperature Inversions Have Cold Bottoms

Dust in the air suspended
Marks the place where a story ended
T. S. ELIOT

*I*t is well known—to those who know it well—that Los Angeles owes its infamous smog to its large number of automobiles and plentiful sunshine for transforming their emissions into noxious substances. It may be less well known that there is a third ingredient in the recipe for smog: the variation of temperature with height in the atmosphere. While Morris Neiburger was senior meteorologist at the Air Pollution Foundation in Los Angeles, he was subjected to a stream of suggestions for modifying the meteorological conditions of the Los Angeles Basin. These suggestions reflected the character of the area: unconventional, imaginative—but unrealistic. Before discussing Neiburger's analysis of the feasibility of one of these meteorological solutions to the smog problem—penetrating or dissipating the inversion—we need to understand what an inversion is and how it is related to the dispersal of pollution.

Demonstrating Atmospheric Stability

A striking demonstration of the connection between atmospheric temperature profiles and pollution was devised by Hans Neuberger and George Nicholas and is included in their *Manual of Lecture Demonstrations*. Place a tall, clear plastic cylinder, tightly sealed at the bottom, in a coffee can partly filled with ice and water. Gently blow smoke—you will need a fair amount—into a tube that passes through a small hole to the bottom of the cylinder. The result is shown in Figure 11.1. Note that the smoke lies

121

Figure 11.1
The bottom of this cylinder is in cold water. As evidenced by the smoke, there is little vertical mixing because warm air overlies cold, denser air. Photograph by Gail Brown.

quietly at the bottom of the cylinder; it does not rise much even if you jostle the cylinder a bit. But now replace the ice water with boiling water. Within a minute the air will suddenly come alive with curling, turbulent motion, ending in thoroughly mixed smoke (see Figure 11.2). Merely changing the temperature at the bottom from low to high dramatically transformed the state of the air in the cylinder: in the first instance it was *stable*; in the second it was *unstable*.

Atmospheric stability is closely connected with buoyancy. Bubbles rise in beer because of buoyancy. Buoyancy also can be demonstrated with a pan of sand and two ping pong balls, one filled with lead shot. If both balls are placed on the surface of the sand and the pan is agitated vigorously, the heavy ball disappears while the light one remains on top. This is shown in Figure 11.3. If the light ball is buried, it will eventually rise to the surface as the sand is agitated; it may be observed to suddenly

122

Figure 11.2
If the cold water in the can shown in Figure 11.1 is replaced by hot water, the air becomes unstable resulting in rapid vertical mixing (left) until the smoke is completely dispersed throughout the cylinder (right). Photograph by Gail Brown.

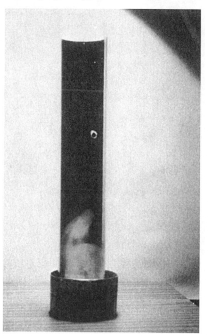 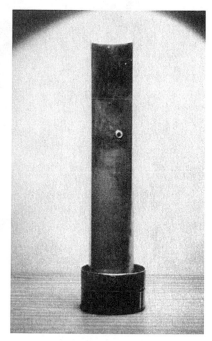

break through the surface and float on the top. The light ball is positively buoyant: it is less dense than the sand; the heavy ball is negatively buoyant: it is more dense than the sand.

Robert Hooke, a seventeenth-century English physicist, used a dish of agitated sand to demonstrate that the ease with which liquids and gases flow—their fluidity—implies that matter is in continual motion. For if you make a hole in the sand with your finger, the hole immediately fills in again. But Hooke was also aware that the behavior of heavy and light objects in agitated sand demonstrates buoyancy. Sir William Bragg, in his superb book *Concerning the Nature of Things*, quotes Hooke as follows: "For by this means [shaking], each sand becomes to have a vibrative or dancing motion, so as no other heavier body can rest on it, unless sustein'd by some other on either side: nor will it suffer any body to be beneath it, unless it be heavier than itself."

Figure 11.3
When the sand in the pan is agitated, the light ball remains on the surface (top) while the heavy ball begins to sink (bottom) and eventually disappears (next page).

Temperature Differences

Cold air is denser than warm air at the same pressure. When the bottom of the cylinder was colder than its top, warm air overlaid cold air. This is stable: if warm air were to sink into the cold air it would be buoyed upward, back whence it came; on the other hand, if cold air were to rise into the warm air it would be heavier than its surroundings and would sink. The reverse—cold air above warm—is unstable: cold air sinking into warm air would find itself colder, hence denser, than its surroundings, and would continue to sink; warm air rising into cold air would be lighter than its surroundings and would continue to be impelled upward by buoyancy. Mixing of layers of air at different temperatures is therefore suppressed if the temperature increases with height. Conversely, mixing is promoted if the temperature decreases with height. Rising air cools, however, and it is the *difference* in temperatures between this air and its surroundings that determines if it will continue to rise or will sink. So we must ask, at what rate must the temperature decrease with height if the air is to be unstable?

Consider a parcel of air, which we may imagine to be enclosed within an elastic balloon (Figure 11.4). If the balloon is impelled upward by some means, it will expand: pressure decreases with

Figure 11.4
Two hypothetical vertical temperature profiles (solid lines). The dashed line shows the dry adiabatic lapse rate.

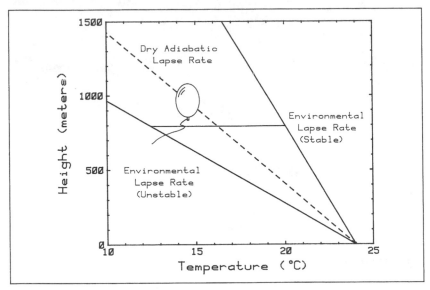

height, and at each instant the balloon increases its volume so that the pressure inside balances that outside. But it takes energy for the balloon to expand against its surroundings, just as it takes energy to lift a weight against gravity. The energy required for expansion is taken from the thermal energy of the air in the balloon, so it cools. There may be water vapor in the parcel, but we assume that it does not condense. And we also assume that the parcel is neither heated nor cooled by its surroundings because of temperature differences between the two. With these restrictions, the rate at which the temperature decreases is called the *dry adiabatic lapse rate*: dry because there is no condensation, adiabatic because there is no heat transfer, and lapse because there is a decrease.

Now suppose that the temperature of the atmosphere decreases with height according to the dry adiabatic lapse rate, which is about 10°C per kilometer. At each height, a rising parcel of air—we may imagine it to be in a balloon carried aloft on the tail of a kite—has the same temperature and pressure, hence density, as that of its environment. If the balloon is released at any height, it will neither sink nor rise.

126

If you move away from a warm stove, the temperature around you decreases. Similarly, the temperature usually decreases with height in the atmosphere, the earth's surface warmed by the sun taking the place of the stove. But the rate of temperature decrease need not be, indeed rarely is, the dry adiabatic lapse rate: it may be greater or less. Suppose that the temperature of the balloon's surroundings decreases more rapidly than 10°C per kilometer; this is shown by the leftmost curve in Figure 11.4. At any point in the ballon's ascent, its temperature, which follows the dry adiabatic lapse rate, is greater than that of its surroundings; if it were released at, say, 1,000 meters, it would continue to rise. In this instance the atmosphere is unstable. But if the temperature decreases less rapidly than 10°C per kilometer (shown by the rightmost curve in Figure 11.4), the temperature of the balloon is always less than that of its surroundings. If released it would therefore sink. In this instance the atmosphere is stable. So the dry adiabatic lapse rate merely defines the boundary between a stable and an unstable atmosphere. If we replace the imaginary balloon with real parcels of polluted air, it follows that, in the context of pollution dispersal, instability is desirable whereas stability is not. We are now better able to understand the meteorological factor in Los Angeles smog.

Los Angeles Smog

The average variation of temperature with height at 7 A.M. in September at Long Beach, California, as taken from Morris Neiburger's article "Weather Modification and Smog," is shown in Figure 11.5; the dashed line indicates the dry adiabatic lapse rate. Below about 475 meters the atmosphere is stable, but only slightly. Between 475 and 1,055 meters, however, temperature *increases* with height, which indicates a very stable atmosphere. This is called a temperature *inversion*, and polluted air under an inversion is trapped there; the temperature inversion acts as a lid on the upward dispersal of pollution. To eliminate the smog, one need merely eliminate the inversion by, for example, heating all the air below its top. Neiburger calculated that to do this over the Los Angeles Basin would require the energy obtained from burning (at 100 percent efficiency) 1.27 million tons of oil, the amount of crude oil processed in 12 days by all the refineries in

Figure 11.5
The average temperature profile at 7 A.M. in September at Long Beach, California (from M. Neiburger, "Weather Modification and Smog," *Science,* Vol. 126, 1957, p. 637). The dashed line indicates the dry adiabatic lapse rate.

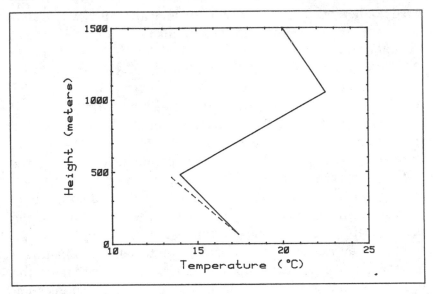

the Los Angeles Basin. And the air would not just sit still while being heated but would rise and be replaced by colder air, thereby reestablishing the inversion.

The prevailing temperature inversion in the Los Angeles Basin, particularly during the warmer months, is an inevitable consequence of the general circulation of the atmosphere. To eliminate the inversion would require either modifying the general circulation or moving the entire basin to a more meteorologically benign location. But neither solution has yet found advocates among even the most fanciful of Southern Californians.

Excuse My Dust

My most vivid—and unpleasant—memories of the consequences of temperature inversions are from my student days in Arizona. I lived on the desert, about 15 miles from Tucson. I would ride my bicycle to the university early in the morning, following less-traveled dirt roads as far as possible to avoid automobiles. Other than the large black dog who would wait patiently for me in hopes of catching me as if I were a jackrabbit (I always won the race by a margin of the dog's hot breath), what I dreaded most was that a car would precede me in the early morning hours. Because of overnight radiational cooling of the ground in the clear desert air, the air near the surface would be colder than that aloft. Dust raised by a passing automobile would hang in this stable air, and I would have to eat the gritty stuff for miles. But at the end of the day, after the sun had been baking the ground for hours, air temperature decreased with height, the atmosphere was unstable, and dust raised by passing automobiles was so quickly dispersed that I hardly noticed it.

*12

Water Vapor Mysticism

The sun was warm but the wind was chill.
You know how it is with an April day
When the sun is out and the wind is still,
You're one month on in the middle of May.
But if you so much as dare to speak,
A cloud comes over the sunlit peak,
And you're two months back in the middle of
March.
ROBERT FROST

I have spent much of my life in a mild or warm climate, about 12 years of it in southern Arizona. For the past 10 years, however, I have lived in Pennsylvania. When I encounter old friends, conversation often turns to the differences between where I used to live and where I live now. Whether in sympathy or in disapproval of my defection from the West, my friends can be counted on to note, shivering reflexively, how miserably cold Pennsylvania winters must be. I respond that even in the arid Southwest, temperatures can be low in winter. "Ah yes," they'll rebound smugly, "but you have wet cold, whereas we have dry cold." I smile to myself when I hear this mix of sense and nonsense but say nothing, not wanting to demolish their sense of superiority for having the good fortune to live in dry cold or their sympathy for my having to endure wet cold.

Not long ago, however, I discovered that the supposed difference between wet and dry cold has been given a physical explanation in a textbook: "A cold, damp day often feels colder than a cold 'dry' one because moist air conducts heat away from the body better than 'dry' air." This has survived into a third edition, evidence that either its author's customers are remarkably credulous, or he is deaf to their criticism. The alleged higher conductivity of moist air is not mentioned as a mere aside, it is

131

set off in a box to indicate its status as a great truth. And yet I state without hesitation that, box or no, this statement about the conductivity of air contains not an atom of truth. To quote the late Wolfgang Pauli, "It isn't even wrong."

Common Sense and Cooling Curves

Water vapor usually makes up less than a few percent of atmospheric air, especially cold air. Thus for water vapor to appreciably affect the thermal conductivity of air—a mixture of various gases and water vapor—each water molecule must be at least a hundredfold more effective at transporting energy than nitrogen and oxygen, the preponderant molecules in air. This necessary heroic effectiveness of water vapor runs counter to common sense. Yet there are examples where small amounts of a substance can give large, even catastrophic, effects. The toxin responsible for botulism is one example that comes to mind. So perhaps the best way to settle questions about the conductivity of moist air is by doing some experiments. I'll return to scientific common sense later.

I discussed cooling curves in Chapter 7. They are easy to obtain, although to do so requires considerable time. Generating them is about as exciting as watching grass grow, but they are instructive. One simply heats water in a flask and then allows it to cool, noting the temperature of the water at various times. I call the time required for the temperature difference between the water and its surroundings to decrease to half its initial value the *cooling time*. The greater the cooling time, the more slowly the flask cools under given conditions.

When I decided to measure cooling times in dry and moist air, my Sri Lankan niece, Manoja Dayawansa, was home with us for the Christmas holidays. I needed her bathroom as a laboratory. She is a budding scientist, so I sparked her interest in my project as a means of getting the use of her bathroom; I even enlisted her as an assistant.

We tried various methods for obtaining different vapor densities (water vapor molecules per unit volume) in the bathroom with its door shut tightly. First we used a humidifier, then a dehumidifier. We gave up on the dehumidifier because, although it decreased the relative humidity, it raised the air temperature, with little net change in the vapor density.

Air temperatures in the bathroom ranged a few degrees on either side of 20°C (68°F). We filled a 250 ml flask with boiling water, then took readings of its temperature as it cooled from about 85°C (185°F). After the humidifier had been running at its highest setting for two hours, the relative humidity in the bathroom was about 100 percent. The vapor density, not the relative humidity, determines the conductivity of moist air. Knowing the air temperature and relative humidity, we could calculate the vapor density.

For these experiments, the vapor density in moist air was more than 50 percent greater than that in dry air. Yet the cooling times were almost identical: 45 minutes in moist air, 45.6 minutes in dry air. This difference is insignificant given the errors in our measurements and their reproducibility. An additional run in moist air gave a cooling time of 45.3 min. The errors in cooling times are at least a few percent. So we concluded that, within experimental error, there is no significant difference between cooling of a body in dry air and in moist air. But I must qualify this by saying that the body itself cannot be wet (although filled with water, the flask was not wet in these experiments). When the body is wet, cooling times can depend markedly on the moisture content of the air.

Evaporative Cooling

When I encounter statements about the alleged great cooling power of moist air, I think to myself that I would expect cooling to be greater in dry air, all else being equal (an important qualification, as we shall see). Evaporative cooling of a wet object is driven by a difference between the density of water vapor at the surface of the object and that in the surrounding air. Even when our skin is dry, our lungs are moist. We inhale air at a temperature lower than our deep body temperature, and the water vapor density in this air is less than that at the surface of our lungs. To breathe is therefore to cool evaporatively. The physical comfort of those who do not breathe is of little concern.

To give some quantitative support to these assertions about evaporative cooling, Manoja measured cooling curves for a flask wrapped with wet cloth (Figure 12.1) in both moist and dry air. The results are shown in Figure 12.2.

133

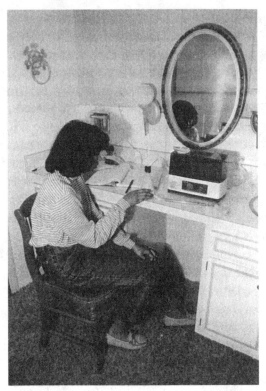

Figure 12.1
As evidenced by the persistence of the cloud from the humidifier, the relative humidity in this bathroom is almost 100 percent. Manoja is measuring the rate of cooling of a flask filled with hot water and wrapped with wet cloth.

For reasons I hope will become apparent, the initial temperature of the flask was taken to be about half that in the previous experiments. The vapor density in the moist air was about 40 percent greater than that in the dry air. Cooling times varied in about the same proportion, that in moist air being about 50 percent greater than that in dry air.

One might be tempted to conclude from these experiments, and from the common experience of being uncomfortable in hot, humid weather, that in cold weather increased water vapor in the air would help to suppress evaporative cooling and we consequently would be more comfortable. This is qualitatively correct, but quantitatively insignificant. Although high humidity makes us uncomfortable in hot weather, the converse is not true.

The reason for this is made clear by inspection of the saturation vapor pressure curve in Figure 12.3. As I stated previously, evaporative cooling depends on the difference in the vapor densities at the surface of the evaporating body and in the surround-

Figure 12.2
The rate of cooling of a 250 ml flask wrapped in wet cloth and filled with hot water depends on the humidity of the surrounding air. The cooling time in a moist environment is about 50 percent greater than it is in a dry environment.

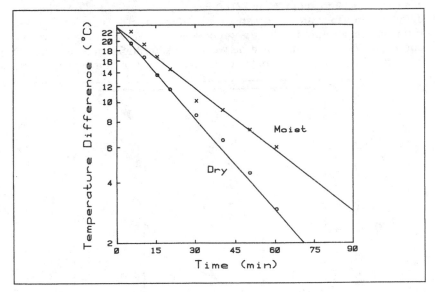

ing air. For our purposes, we may take vapor pressure to be synonymous with vapor density, which is approximately true over small absolute temperature ranges.

Consider first the difference between the saturation vapor pressure of water at about 37°C (98.6°F), the deep body temperature, and the vapor pressure in air at a temperature of 26°C (79°F). Depending on the relative humidity of this air, the vapor pressure difference can vary by a factor of almost two. But now consider air at 0°C. The difference between the vapor pressure of this air and the saturation vapor pressure at 37°C varies hardly at all when the relative humidity ranges from 0 to 100 percent. Thus high relative humdities can make us uncomfortable in hot weather, but in cold weather humidity only negligibly affects our comfort. Evaporative cooling can change appreciably with relative humidity only when the evaporating body and the surrounding air have similar temperatures. This occurs for humans in hot weather but not in cold.

There were limits to how cold we could make Manoja's bathroom, so we went in the opposite direction: we measured cooling

Figure 12.3
Evaporative cooling is driven by a difference between the saturation vapor pressure of the evaporating water and the vapor pressure in the surrounding air. At 37°C, the deep body temperature, the saturation vapor pressure is about 63 mb. If the air temperature is 26°C, the vapor pressure in the air can vary from 0 to about 34 mb (100 percent relative humidity). Thus the difference in vapor pressures can vary by more than a factor of two. But if the air temperature is 0°C, the vapor pressure can vary only between 0 and 6 mb. Thus the difference between the saturation vapor pressure at 37°C and the vapor pressure of the colder air can vary by at most 10 percent.

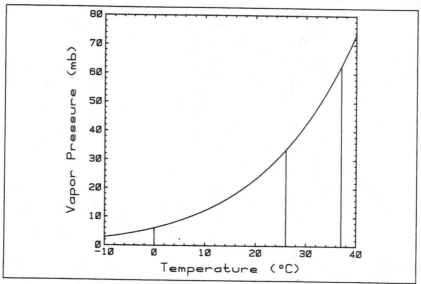

curves for an initially hot (90°C), wet flask in moist and dry air at temperatures around 20°C. The difference in the cooling times was small.

Molecular Interpretation of Gas Conductivity

Simple physical reasoning leads to the conclusion that the thermal conductivity of air is essentially independent of its water vapor content.

In a gas to all appearances at rest, molecules are whizzing about randomly in all directions. When molecules move, they carry with them their properties: mass, momentum, energy. The average kinetic energy—energy of motion—of gas molecules increases with temperature. If the temperature of a gas is the same everywhere, there is no net transport of molecular kinetic energy from one region to another. Only when there are temperature differences is there a net transport of energy. The ratio of the rate of energy transport to the temperature gradient (the rate of temperature change with distance) is called the *thermal conductivity*; the greater the thermal conductivity, the greater the rate of energy transport for a given temperature gradient.

The rate at which molecules cross a small area in a gas depends on their number density and how fast they are going. The number density is proportional to pressure and inversely proportional to absolute temperature; the mean speed of molecules is proportional to the square root of absolute temperature and inversely proportional to the square root of their mass (this is because the kinetic energy of a molecule depends directly on its mass and the square of its speed). So the flux of molecules is directly proportional to pressure and inversely proportional to the square roots of both absolute temperature and mass.

Molecules cannot transport their energy from one region to another unless they collide with other molecules. The average distance a molecule travels between collisions is called its *mean free path*. The greater the density of molecules and the larger their cross-sectional area, the shorter the mean free path.

Consider two regions of a gas one mean free path apart. The temperature difference between them is the temperature gradient times the mean free path. Thus the net energy transport from the higher temperature region to the lower is proportional to the temperature gradient, the mean free path, and the molecular flux. This flux and the mean free path depend on pressure in opposite ways, the first increasing with pressure, the second decreasing. So we arrive at the perhaps surprising result that the conductivity of a gas is independent of its pressure, although it does increase with the square root of its absolute temperature.

For our purposes here, what is of greater interest is that the conductivity of a gas depends inversely on the square root of the mass of its molecules and their cross-sectional area. The mass of a water molecule and its size are not greatly different from those of nitrogen and oxygen. So we must conclude from our crude arguments that the thermal conductivity of water vapor cannot be much greater than that of dry air. In fact, the conductivity of water vapor is slightly *less* than that of either nitrogen or oxygen, although this is not explicable by simple arguments.

137

To determine the conductivity of a mixture of gases such as air from the conductivities of its constituents is not trivial. But it is approximately true that the conductivity of a mixture is a weighted sum of the conductivities of its constituents. Thus, all else being equal, the conductivity of moist air should be somewhat *less* than that of dry air, although the difference is so small as to be of no consequence to human comfort.

The independence of the conductivity of air on pressure should put to rest the spurious explanation of why we feel colder on high mountains: the air is "thinner" there. The conductivity of air does depend on the square root of absolute temperature, so in going from Death Valley to the top of Mt. Everest, the conductivity decreases by perhaps 10 percent. But this has essentially nothing to do with why we usually feel colder on mountains than on deserts. Indeed, on the basis of conductivity arguments alone, we should be warmer on the summits of high mountains.

Origins of a Fallacy

The fallacy that the conductivity of moist air is sufficiently greater (in fact, it is smaller) than that of dry air to affect human comfort perhaps has its origins in the confusion between water vapor and liquid water. I do not doubt that the conductivity of liquid water is vastly greater than that of air, moist or dry, and that this has profound consequences: we can be immersed in air at 5°C (40°F) without much discomfort, yet in water at this temperature we should soon die. But it is a giant leap from liquid water to moist air. The number density of water molecules decreases by a factor of 100,000 in going from the one to the other, and it is unreasonable to expect that with this huge change, arguments valid for one substance will be valid for the other.

The commonly held view that one is generally more uncomfortable in a cold, moist climate than in a cold, dry one is not incorrect. What is incorrect are attempts to explain this by specious physical arguments. The author of the poem at the head of this chapter had a better grasp of the determinants of human comfort than the author of the textbook I cited. A moist climate means rain. Rain means clouds. Clouds mean obscuration of the sun, hence reduced sunshine. Our physical comfort depends as much on sunshine as on air temperature.

I spend as much time as possible out of doors in all kinds of weather. On a cold but windless and sunny day I can be moderately comfortable even wearing a light shirt. But let a cloud pass between me and the sun and I must reach into my knapsack for a parka or a sweater, reminded of these lines from a favorite Shakespearean sonnet:

> Even so my sun one early morn did shine
> With all-triumphant splendor on my brow;
> But out, alack! he was but one hour mine;
> The region cloud hath mask'd him from me now.

Water Vapor Mysticism

The alleged high conductivity of moist air is just one example among many of what might be called water vapor mysticism: the assigning of all kinds of miraculous powers and properties to water vapor. Meteorologists are perhaps the most active propagators of water vapor mysticism. They know that water and its transformations play a great role in the running of earth's weather machine. But from this it does not follow that water vapor is responsible for everything in the atmosphere.

Mirages (see Chapter 6) are sometimes attributed to water vapor. The blue sky is sometimes attributed to scattering by water molecules, as if nitrogen and oxygen were powerless to scatter light. Yet the sky would be just as blue and mirages wouldn't be any different if the atmosphere were bone dry.

It is indeed true that water is a fascinating substance and is responsible for all kinds of processes. Without water in its various forms, life would not likely be possible. Yet water vapor does not do everything. It is not *the* universal solvent—except perhaps for ignorance.

⁎ 13

Strange Footprints in Snow

*We have found a strange footprint on
the shores of the unknown. We have
devised profound theories ... to
account for its origin.*
Arthur S. Eddington

*O*ne morning I was tramping along a hillside trail on my
way to catch a bus in a nearby town. A light snow had
fallen the previous night. This trail is one I travel often, so my
feet can find their way over it without much conscious guidance.
On a familiar trail I can safely daydream, which is what I was
doing that crisp morning. Suddenly I was shaken from my trance
by a glowing footprint in the snow (Figure 13.1), left there by
my wife striding ahead of me. How can this be? I thought. My
eyes presented me with a scene that my mind thought to be
impossible: you can't make snow brighter by stepping on it.
When faced with an apparent contradiction between what is and
what at first blush you think ought to be, more careful thought
is in order.

The Invariance of Optical Thickness

Snow on the ground, like a cloud in the sky, is a multiple-scattering medium: each scatterer—ice grain or water droplet—is illuminated not only by incident sunlight but by light scattered
by all its neighbors as well. Grains in snow are much more
densely packed than droplets in a cloud and also much larger,
but otherwise the two are similar. A whiteout, in which all con-

141

Figure 13.1
Why is this footprint so much brighter than its surroundings?

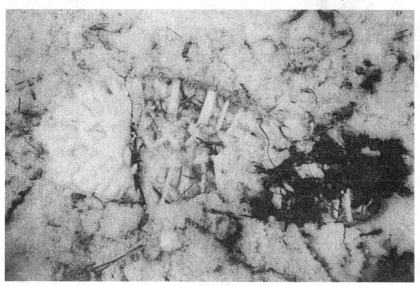

trast between snow and fog disappears, is evidence that snow and clouds can be optically indistinguishable.

One of the most important determinants of the brightness of any multiple-scattering medium is its *optical* thickness. This is not the same as its *physical* thickness, although the two are related: optical thickness is physical thickness measured in units of mean free path, the average distance light propagates in the medium before it is scattered (or absorbed, although for our purposes we may ignore absorption). Optical thickness is therefore proportional to physical thickness, where the proportionality factor contains the number density of scatterers. The greater this density, the shorter the scattering mean free path.

The optical thickness of a cloud (or a snowpack) is unaffected by compaction because the number density of scatterers increases in the same proportion as the physical thickness decreases. And similarly for expansion, the density of scatterers decreases in the same proportion as the physical thickness increases. Because the product of number density and physical thickness is constant, the brightness of an illuminated cloud should not change when compacted provided that its droplets are not brought so close together that they coalesce into larger ones. And the brightness

Figure 13.2

These two suspensions of scattering particles (represented by small circles) are identical seen from above (top). They present the same number of scatterers to a unit area of an incident beam even though the suspension on the left is physically twice as thick as that on the right. As seen from the side (bottom), the suspension on the right presents more scatterers to a unit area of an incident beam.

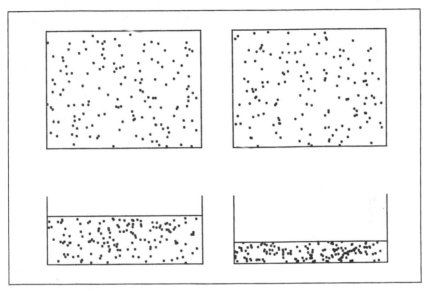

of a snowpack should not change when compacted provided that one stops well short of forming a slab of ice.

Within limits, therefore, we expect the brightness of a cloud or snow to be unaffected by compaction (or expansion). To drive this point home, I prepared the schematic diagram in Figure 13.2, which shows plan and elevation views of two suspensions of scatterers. Although the total number of scatterers is the same for both suspensions, one is twice as thick (physically but not optically) as the other. Seen from above, however, they are identical. Light incident on either suspension confronts the same number of scatterers. As far as this light is concerned, the two are identical.

Perhaps now you can better understand why at first I was so puzzled by the bright footprint: I knew—or thought I knew—that compacting snow cannot make it brighter. And my expectations were based on more than theory. Many years before, my colleague Bob Beschta and I had measured what happens to the

143

Figure 13.3
Why is this footprint, unlike that shown in Figure 13.1, only as bright as its surroundings?

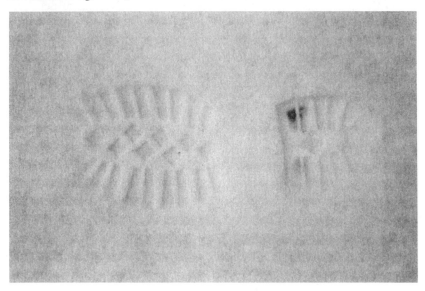

reflectance of snow when it is compacted. At the time, we were both at the University of Arizona. One winter, in Flagstaff, we used a snowmobile to compact fresh, deep, powder snow. We set up two radiometers, one to measure incoming solar radiation and one to measure solar radiation reflected by the snow. We first measured the ratio of the reflected radiation to that incident for the undisturbed snow. Then we roared over this same snow with the snowmobile, increasing the density of the snow by as much as a factor of two. We could detect no difference: the compacted snow reflected the same fraction of solar radiation as the powder snow. Our data showed variations, but none that pointed to a systematically different reflectance for snow of one density than for that of another.

If further proof is necessary, you'll find it in Figure 13.3, a footprint in snow showing no brightness difference between the compacted snow and its undisturbed surroundings. But now I seem to have piled a contradiction onto a mystery: first I showed you a footprint that is brighter than its surroundings, then I showed you one that is not. It is this apparent contradiction, however, that leads to the solution of the mystery.

A Mystery Solved

Although I have walked through snow many times, often for long distances, I cannot recall ever having been startled by bright footprints in snow until that winter morning not long ago. What was so special about this one event?

My wife was wearing boots with thick lug soles, the snow was somewhat wet, and only a thin coating of snow had fallen, although more than just a dusting. All three of these ingredients were necessary to yield what I observed.

Snow was compacted by my wife's boots, and it was cohesive enough to accumulate in the ample spaces between lugs. From time to time, a slab of this compacted snow would become dislodged from a boot and fall onto the ground. What I was seeing was therefore not really footprints in snow, but rather casts in snow of the boot's sole, several accumulated footprints. The optical thickness of the snow in a cast was greater than that of the surrounding snowpack, enough to give noticeable contrast between the two. Not much snow need fall before a snowpack becomes optically thick, that is, effectively infinitely thick. Had the snow been much thicker, the contrast between it and a cast would not have been enough to catch my attention. But if the snow had been thinner, not much snow would have accumulated between the lugs. The snow also had to be cohesive. If it had been loose powder snow, not much would have accumulated. And, finally, had my wife been wearing smooth-soled shoes or boots, she would have shed no (relatively) bright snow casts.

A Few Experiments and Another Mystery

In the course of unraveling the mystery of the bright footprint, I devised a few experiments using aluminum pans painted black, filled to different depths with suspensions of watered milk of various concentrations, and illuminated from above.

The first experiment is shown in Figure 13.4. I kept the concentration of the milk suspension constant and varied its physical, hence optical, thickness. Not surprisingly, the brightness of the thicker suspension was greater. This exemplifies a general rule: the brightness (for given illumination) of a multiple-scattering medium seen from the side of the illuminant always increases with optical thickness.

Figure 13.4
Both blackened pans contain milk diluted with water to the same concentration. The physically thicker, hence optically thicker, suspension on the left is considerably brighter than the suspension on the right.

In the second experiment, I poured the same amount of watered milk into both pans. Then I added pure water to one of them to increase the depth of the suspension. This is equivalent to expanding a cloud: the concentration of scatterers in the medium decreases as its physical thickness increases, the product of the two remaining the same. Note that the two suspensions shown in Figure 13.5 appear almost the same, although the physically thicker suspension is slightly darker. The invariance principle I invoked previously is strictly valid only if the medium is infinite in lateral extent. No medium is literally infinite, thus *infinite* here just means so thick as to be effectively infinite: an increase in lateral dimensions would result in no perceptible change.

What if the medium is laterally thin? The answer to this question follows from contemplating the diagram in Figure 13.2. Both suspensions have identical *normal* optical thicknesses, but their lateral or *transverse* optical thicknesses are different. That is, in plan view the suspensions are identical, whereas in elevation view they are not. The transverse optical thickness of the physically thicker suspension is less than that of the thinner one.

Figure 13.5
These two suspensions are about equally bright, even though that on the right is physically twice as thick as that on the left. The suspension on the right is slightly darker, especially near its edges, because although the normal optical thicknesses are the same for both suspensions, the transverse optical thicknesses are not.

Figure 13.6
These two suspensions are more dilute than those shown in Figure 13.5. The normal optical thicknesses of the two are the same, but the transverse optical thickness of the dark suspension on the left is less than that of the brighter suspension on the right.

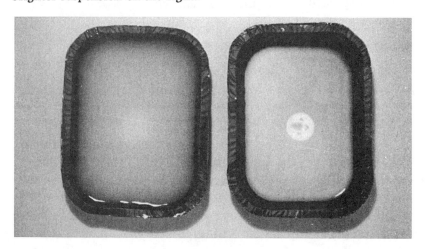

As a consequence, light diffusing laterally through the thicker suspension is more likely to reach boundaries where it is absorbed, thus not contributing to the brightness as seen from above.

To demonstrate this point, I did an experiment similar to the previous one but used more dilute milk suspensions. I filled two pans to the same depth, then added water to one of them. The brightness difference between the two suspensions was now striking (see Figure 13.6). One can see something like this in clouds. Given two small clouds, identical in vertical thickness, drop size and concentration, and illumination, the laterally smaller cloud will be less bright.

To demonstrate the thinness of the suspensions used in my final experiment, I placed bus tokens in the bottom of each pan. This presented me with something I had not forseen: the token in the brighter suspension was more distinct. Had I been asked to make a prediction about this beforehand, I would have predicted the reverse on the basis of my musings about the following question: What is the least thickness a cloud can have before you cannot see the sun through it? (See Figure 13.7.) I had always considered this to be a problem of contrast. One can no longer see the sun when its unattenuated light is as bright (within the contrast threshold) as the surrounding diffuse cloudlight, the brightness of which increases with cloud thickness (up to a point). Because the sun becomes indistinct as its surroundings become brighter, I would have expected the bus token in the brighter suspension to be more indistinct.

The Nude-in-the-Shower Phenomenon

Not long after I had stumbled on the puzzle of the bus tokens, I gave a talk at the Kerr-McGee Technical Center in Oklahoma City. After the main course, I served this puzzle as dessert. Afterwards, Tom Cawthon, the director of the center, showed me how to solve it. As I watched him sketching at the blackboard, I realized suddenly that what I had observed was a variation on what David Miller and George Benedek, in their book *Intraocular Light Scattering*, call the nude-in-the-shower phenomenon (see Figure 13.8). Two phenomena that I previously had kept in separate mental compartments began to merge.

Figure 13.7
Scattering by water droplets reduces contrast between the sun and its surroundings. A cloud only about 50 meters thick obscures the sun completely.

The distinctness of an object decreases as it is moved farther from a diffusing screen, such as the frosted glass door of a shower, or a strip of what is called transparent (but should be called translucent) adhesive tape, or a sheet of "nonreflecting" glass. To understand why, consider the diagram in Figure 13.9. A point source at *P* sends light to an observer at *O*. When a diffusing screen is interposed between observer and source, light from it illuminates the screen, which scatters light in all directions. Thus the observer receives light from directions he previously had not: the point source is now surrounded by a halo of scattered light. What is observed would be no different if this halo were in a plane containing *P* and parallel to the screen.

Scattering by particles larger than the wavelength is strongly peaked in the forward direction. We may consider a diffusing screen to be simply a piece of smooth glass covered with a dusting of such particles. The angle *a* at which the scattered intensity drops to, say, half the value of that scattered in the forward

149

Figure 13.8
The clarity of details of an object seen through a diffusing screen depends on the distance between them.

direction, is small. Thus the halo radius r, arbitrarily defined as the distance at which the intensity is half that of the point source, is proportional to the angle a and the distance d between source and screen.

To an observer, the effect of a diffusing screen is to surround a point source with a diffuse halo of light centered on it, the intensity of which decreases with distance from the center. The distance r at which the intensity falls to half the central value depends on the details of scattering (e.g., the angle a) and the distance from source to screen. Suppose that a second point source were to be placed in the plane of P at a distance r from it. Because of the screen, the contrast between these two sources would be reduced, since their halos overlap. Without the screen in place, the two sources would be distinct; with the screen, they are less distinct. A diffusing screen therefore effaces details of an object to an extent that depends on its distance from the screen. Note that the details of the hand pressed against the shower door

Figure 13.9

Light from the point source *P* is scattered by the diffusing screen. An observer at *O* therefore receives light from directions he would not have otherwise, and the source is surrounded by a halo. If *a* is the scattering angle for which the scattered intensity is half that of the source, the corresponding halo radius *r* is *a* times *d* if *a* is small.

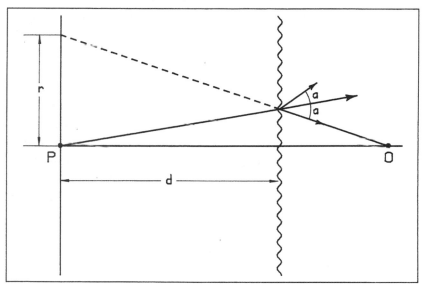

are distinct, but those of the more distant body are not. And you cannot read a book while looking through translucent tape unless it lies almost on the page.

To test Tom Cawthon's suggestion that I had observed a variation on the nude-in-the-shower phenomenon, I took two large pans, the bottoms of barbecue grills, each about one-half meter in diameter, and painted their bottom surfaces with flat black paint and their sides with flat white paint. The purpose of using large pans with white sides was to simulate laterally infinite media because the interpretation of my previous observation had been clouded by the finite lateral dimensions of the containers.

I filled one pan almost to the brim with water, the other less than half way. I placed a bus token in the bottom of each pan. The tokens were polished so that they were identical as far as I could tell. When I added equal amounts of milk to both pans of water, halos appeared around the brass tokens; they looked as if they were lying on beds of gold dust.

Although the physical thicknesses of the two suspensions were unequal, their (normal) optical thicknesses were the same and the consequences of their different transverse optical thicknesses were negligible, which I verified by observation. I looked at both suspensions through paper tubes, one for each eye, to avoid being confused by edges. To within my ability to compare, the suspensions appeared identical. I even switched eyes, first looking at a suspension with one eye, then the other.

Without a doubt, the token seen through the physically thicker suspension was less distinct. I could detect both tokens with about equal facility, but I could discriminate the details of one much better than those of the other.

The interpretation of this is as follows. A milk suspension is analogous to a diffusing screen. Thus we should be able to replace a suspension with such a screen. The distance of the equivalent screen from a token will depend on the physical thickness of the corresponding suspension; the thicker the suspension, the greater this distance. Thus it is no wonder that the token seen through the physically thicker suspension was less distinct.

At the end of the previous section, I posed a question but then passed over it: What is the least thickness a cloud can have before you cannot see the sun through it?

Every time I do a back-of-the-envelope calculation to answer this question, I obtain a somewhat different answer. My best guess to date is that a cloud with an optical thickness of about 10 will just obscure the sun. The corresponding physical thickness will depend on the drop size distribution and liquid water content of the cloud. For typical cloud parameters, an optical thickness of 10 corresponds to a physical thickness of perhaps 50 meters (about 150 feet). This is consistent with crude observations. For example, I took the photograph in Figure 13.7 one foggy morning on my way to pick up the newspaper. My house was not shrouded in fog, but my neighbor's house (180 feet lower in elevation) was. It was there that I photographed the sun appearing and disappearing in the fog.

Except for its rim, the sun is essentially featureless to the unaided eye. But suppose we could see some features on the sun. I daresay that these would become indistinct before the sun would disappear entirely from view. We could still make out a fuzzy sun even though it would be featureless. Scattering affects brightness contrast and image quality, although the two are not completely independent.

The nude-in-the-shower is a single-scattering phenomenon, whereas the disappearing sun is a multiple-scattering phenomenon. But the two can be brought into rough congruence by expressing the one in the language of the other. We may ask, Under what conditions will the angle *a* (see Figure 13.9) be 90 degrees? When *a* has this value, its tangent is infinite, hence so is the radius of the halo around a point source seen through the scattering medium. This radius can be increased without limit in two ways, thereby obscuring all details of an object viewed through a scattering medium: move the object farther from the medium or increase its optical thickness.

☀14

The Doppler Effect

C. DOPPLER TAKES A WALK

HELLO, FRAU GINZLER.

IT IS INDEED GOOD TO SEE YOU...

ON SUCH A FINE DAY.

AND GIVE MY REGARDS...

TO YOUR HUSBAND, OTTO.

Reprinted with permission by Sidney Harris.

Doppler radar is all the rage in meteorology these days. I am told that it is even mentioned on television. Lest *Doppler radar* become yet another bit of unassimilated jargon that everyone can mouth but few understand, I offer the following.

A Train of Thought

T. P. Gill's book *The Doppler Effect* is too advanced to be recommended for the general reader, but its introduction contains insights I have found enlightening. In particular, Gill's definition of the Doppler effect is succinct and clear: "By the Doppler effect is meant the change in the apparent time interval between two events which arises from the motion of an observer together with the finite velocity of transmission of information about the

155

events." He then goes on to suggest an example: "One might think of a Doppler effect arising when letters are posted from successive railway stations on a long train journey." This hint inspired me to develop the following arguments.

Consider a train moving at constant speed on a stretch of straight track alongside of which stations are regularly spaced. At each station, a car is waiting with its engine running. As the train passes a station, someone on board throws a letter to the car's driver, who immediately heads for a distant house to deliver the letter (see Figure 14.1). The time interval between posting letters is constant; it is the distance between stations divided by the train's speed. But what about the interval between the times they are received? Your immediate response might be that these two intervals are equal. You would be correct if the car's speed were infinite, in which instance it would travel a finite distance in zero time. But because of its finite speed, the two time intervals are unequal. An observer on the train drops letters off at a certain frequency, one every fifteen minutes, say, whereas an observer

Figure 14.1
A train moving at constant speed has an observer on board who tosses letters to the drivers of cars waiting at each station. These cars are driven to a distant house where the letters are delivered. Although the frequency of posting letters is constant, that of delivering them is not: it is greater when the train is moving toward the house, less when it is moving away.

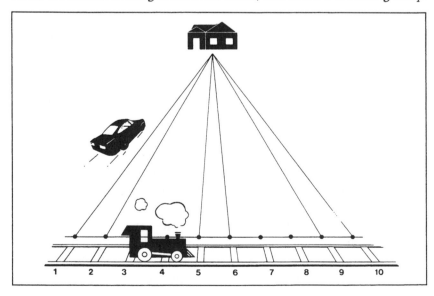

at the house receives them at a different frequency in general. Which frequency is greatest depends on whether the train is moving away from or toward the house. It is easiest to explain why by examining special cases.

Consider first the letters posted at stations 1 and 2 (Figure 14.1). Suppose that the first car arrives at the house just as the train passes the second station. That is, the time it takes the train to go between stations is the time it takes the first car to travel to the house. At the instant the first letter arrives, the second car begins its journey. But this journey is shorter than the first. Hence if the two cars have the same speed, the time between the two deliveries is less than that between the two postings, and the letters are received at a greater frequency than they are sent. Between stations 1 and 2 the train was moving toward the house. What if the train is moving away from it, between stations 9 and 10, for example? Again, assume that the speeds of train and car are such that the ninth car arrives at its destination just as the train passes the tenth station. The journey from this station is longer than that from the ninth. Hence if the two cars have the same speed, the time between the two deliveries is greater than that between the two postings, and the letters are received at a lesser frequency than they are sent. These results imply that when the train is neither moving away from nor toward the house, the frequencies of posting and receiving are the same. That this is so is evident from Figure 14.1, which shows equal distances from stations 5 and 6 to the house.

To make my point as simply as possible, I chose special cases. But my conclusions are valid in general. They can be stated succinctly thus: Signals (letters) transmitted (posted) at a given frequency are received at a different frequency depending on the relative motion of transmitter and receiver. If the transmitter is moving toward the receiver, the frequency of reception is greater than that of transmission; if the transmitter is moving away from the receiver, the frequency of reception is less than that of transmission. The degree to which these two frequencies are different depends on the speed with which the signals are transmitted (the speed of the car for the example chosen). The higher this speed, the closer these frequencies are to each other (in the limit of infinite speed of transmission, they are identical). But notions of what is large (or small) are not absolute. Large (or small) relative to what? It is *ratios* of speeds that determine frequency differences. Such a ratio is the speed at which transmitter and receiver approach or recede divided by the speed of signal transmission; the smaller this ratio, the closer the frequencies of transmission and reception.

If the track were a circle with its center at the house, there would be no difference between the frequencies of transmission and reception of letters regardless of the ratio of the train's speed to that of the car's. In this instance the train is neither moving toward nor away from the house. Stated another way, there is no motion along the line joining transmitter and receiver.

A Smattering of History

Christian Doppler, an Austrian, was the first (1843) to explain the effect that carries his name. He was also the first to misapply it. Doppler recognized that the frequency of a source of sound or light moving relative to an observer must increase or decrease according to whether it is approaching or receding. And he also gave a quantitative relation between the transmitted and received frequencies as a function of the signal velocity and relative velocity of source and observer. He first applied his relation to sound, showing that to raise the pitch of a pure note from C to D requires the source to be moving toward the observer at about 40 meters per second (90 miles per hour). But this was only illustrative. The primary purpose of his paper in the *Proceedings of the Royal Bohemian Society of Learning* is evident from its title: "On the Colored Light of the Double Stars."

Suppose that a star emits a continuous spectrum of light, visible and invisible. If the star is moving away from us, its spectrum will be shifted toward lower frequencies (i.e., toward the red); if it is moving toward us, its spectrum will be shifted toward higher frequencies (i.e., toward the blue). Doppler mistakenly believed that the spectrum of light emitted by stars ends abruptly at the violet and red ends of the spectrum (he seems to have been unaware of light beyond the visible, either ultraviolet or infrared); thus he concluded that a star would become invisible if it were moving at a sufficiently high speed. He did not realize that ultraviolet radiation emitted by a receding star, say, would be shifted into the visible part of the spectrum to take the place, at least partly, of the visible light that had been shifted into the infrared.

Doppler explained incorrectly the colors of stars by invoking an effect he enunciated correctly. He could not have known that the extremely high velocities (appreciable fractions of the speed of light) stars must have for their color to be markedly affected by their motion are rare. Indeed, it is by making use of his effect that the relative velocities of stars are determined, but not in the way he envisioned. My admiration for him is not diminished a whit by his failures, which are eclipsed by his successful exposition of a simple but universal property of signals that had escaped the attention of all his predecessors.

It is fitting that I invoked a train to explain the Doppler effect, for a train played an important role in its experimental verification. This was undertaken by the Dutch meteorologist Buys-Ballot about two years after the appearance of Doppler's paper. Buys-Ballot obtained for a few days the use of a train pulling an open car along a straight stretch of track. The source of sound was trumpets played by skilled musicians, and musicians were also the observers. He alternated sources and observers, placing them in the moving train and by the side of the track. A moving musician would play a given note, and a stationary musician would record its apparent pitch. What was C for the player might be D for his listener (this easily detectable shift does not require astonishing speeds). Buys-Ballot's results confirmed the correctness of Doppler's ideas. When Buys-Ballot first told the musicians about the effect he was trying to verify, they denied that it could exist. They reasoned that the *noise* of approaching trains is no different from that of receding trains. And they were correct: noise is not sound of a single frequency but rather a broad range of frequencies. It is only when a pure note is sounded that changes in pitch can be detected easily. It is to pure notes that we now turn for a demonstration of the Doppler effect.

A Doppler Effect Demonstration

A few years ago Alistair Fraser and I devised a Doppler effect demonstration at the cost of horrifying our colleagues. They had thought they were inured to our inanities until the day we roared up and down the halls of our building carrying blaring instrument tuners.

Figure 14.2
These tuners are used by musicians to determine whether their instruments
are in tune. They also can be used to demonstrate the Doppler effect.

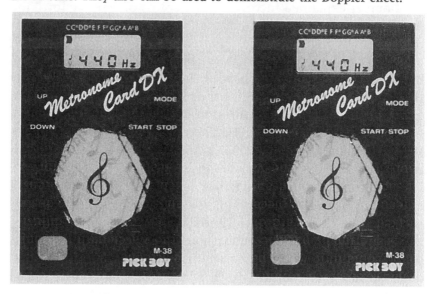

Two tuners, like those shown in Figure 14.2, are required.
We originally used Alistair's bagpipe tuners (an oxymoron if there
ever was one), but subsequently much less expensive tuners have
become available. These tuners produce the notes of the musical
scale.

With these tuners, an exciting classroom demonstration of
the Doppler effect can be obtained at the price of sweat and a
red face (or possibly a heart attack). Set both tuners to the same
note. Then, after having cleared the aisles, grab one of the tuners
and run the length of the room as if pursued by demons. The
students listen for the *beat frequency* and estimate its value, from
which your speed can be estimated.

What exactly is a beat frequency? When two sources of sound
with different frequencies are superposed, the result contains four
components: two with the frequencies of the individual sources,
one with their difference, and one with their sum. The beat frequency is the *difference* of frequencies.

Let me give an example of a beat frequency. One summer,
on a trip west in my 1969 pickup truck, I would hear a low-
frequency hum at certain speeds. This unnerved me because the

more than 170,000 miles on the truck's odometer kept me ever alert for the precursors of catastrophic failure. The frequency of the hum was so low that I could estimate it, perhaps a few hertz (the hertz, abbreviated Hz, is the unit of frequency: one hertz is one cycle per second). The rotational frequency of the engine at 45 mph is perhaps 50 Hz (i.e., 3000 rpm); that of the wheels about a fourth of this. Both of these frequencies were appreciably higher than what I was hearing, which puzzled me. When I got to San Francisco I discovered the source of the hum. Because of worn kingpins, each of the two front tires was scalloped, and the scalloping was not quite the same on both. Hence the frequencies of the sound generated by them were slightly different. The low-frequency hum I was hearing was the difference between these two frequencies, the beat frequency.

In the classroom it is easy for students—at least most students—to hear the beat frequency as you run with tuner in hand. It is a low-frequency tone superposed on the much higher frequency of the tuner. To help the class measure the runner's speed, it is convenient to pick a particular frequency for the pure note emitted by the tuner. The speed of sound is about 330 meters per second. The fractional shift in frequency of a moving source is approximately the ratio of its speed to that of sound. Hence, if you set the tuner to emit sound of frequency 330 Hz (close to E above middle C), the beat frequency gives the runner's speed directly in meters per second. When I have done this demonstration in class, we have measured beat frequencies of about 5 Hz, which corresponds to a speed of 5 meters per second (11 mph). This is a reasonable figure. Olympic sprinters can average twice this over much greater distances than the length of a classroom.

This classroom demonstration has been quite successful, perhaps as much for its scientific content as for the spectacle it presents of a middle-aged, sweating, red-faced professor who—so the students can hope—just might keel over from his exertions. When I am in an especially theatrical mood, I take off my shoes and run in stocking feet.

A Misconception Dispelled

The cartoon by Sidney Harris at the head of this chapter gave me a chuckle when I first glanced at it. But it embodies a misconception, namely, that the apparent frequency of a moving source of sound goes through a maximum (or minimum) rather than changing steadily from high to low.

Given that Doppler is the character in this cartoon, the changing boldness of the letters in the balloons was surely meant to convey a changing frequency (or pitch), not intensity, of the sound received from an object as it moves past an observer. I'll assume that the bolder the letters, the lower the pitch (although it is largely irrelevant if the bolder letters were intended to convey higher pitches); the steady rise of the position of the words in the balloons is consistent with this assumption. The cartoon therefore shows a rising, then falling, pitch as Doppler passes by Frau Ginzler. In reality, the pitch only falls steadily. For example, Figure 14.3 shows the pitch perceived by a stationary observer as a 125-mph train (common in Britain but not, alas, in the United States) with its whistle blowing approaches and then recedes.

To an observer standing by the tracks, the apparent whistle frequency is shifted by an amount proportional to the component

Figure 14.3
The pitch heard by a stationary observer from a whistle on a 125-mph train changes from high to low as the train approaches and then recedes. Simultaneously, the whistle grows louder as the train approaches, reaches a maximum when the train is closest to the observer, then fades as the train recedes into the distance.

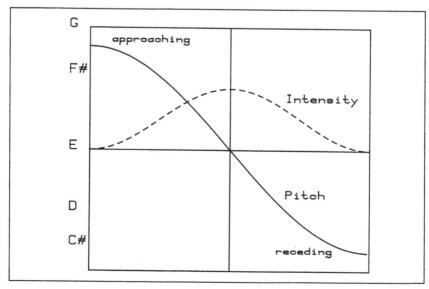

of the train's velocity along the line between it and the observer. When the train is far away and moving toward the observer, the frequency he hears is greater than that heard by the engineer. As the train moves closer, the apparent whistle frequency decreases. Just as the train passes the observer, both he and the engineer hear the same frequency. As the train recedes, the apparent frequency heard by the observer, now less than that heard by the engineer, continues to decrease. Thus a plot of frequency heard by the observer as a function of time shows a steadily decreasing value rather than, as the cartoon implies, a transition from low to high back to low (or from high to low back to high).

Many students think that the pitch of the sound from a moving object rises, then falls (or vice versa). My guess is that the origin of this misconception lies in the changing intensity, which is also shown in Figure 14.3. It seems fairly easy to be confused about changing pitch when intensity is also changing. The cartoon correctly conveys the changing intensity (if we take higher, less-bold letters to indicate greater intensity) but not the changing pitch of Doppler's greetings to Frau Ginzler as he strolls past her.

Doppler Radar

Our demonstration of the Doppler effect makes use of sound waves. Doppler radar relies on electromagnetic waves. My first encounter with Doppler radar occurred when I was a student at the University of Arizona. From the newspapers, I had learned about a new gadget used by the police to catch speeders, but I had never seen one. Then one day, while riding my bicycle to the university, I saw a policeman with a strange object in his hand. So I stopped, approached him, and asked if his gadget was what I thought it was. He was quite proud of it and needed no coaxing to give me a demonstration. "Do you want to see how it works?" he grinned. "See that car coming toward us? Point this at the car, and its speed appears on this display. Oops, I'll see you later." The car was speeding, so off he went after it.

Because the speed of all electromagnetic waves (that of light) is about a million times that of sound, the Doppler effect for light is proportionately less than for sound. Thus to measure my running speed by the change in frequency of electromagnetic waves requires measuring beat frequencies much less than a millionth of the frequency of the radiation used. With modern electronics this can be done readily.

A Doppler radar can measure the motions of storms along the direction of the beam its antenna transmits. Scatterers in the atmosphere (raindrops, hail, snowflakes, insects, etc.) are illuminated by this beam. Because they are moved by the wind, the frequency of the radiation illuminating them is slightly different from that transmitted. They scatter radiation of this shifted frequency back to the antenna. Because of the relative motion of the scatterers (which themselves can be looked upon as small transmitters) the radiation received by the antenna is shifted in frequency again. The relative difference between the transmitted and received frequencies is equal to twice the speed (along the beam) of the scatterers divided by the speed of light.

Doppler radar can determine wind speeds within a radius of only about 50 km. Such radars are expensive, so dense national or international networks of them are out of the question. During the past few years, however, a project (to which I am contributing in a small way) to measure the global winds from satellites has been hatching. This project is simple in conception (although it may not be in execution): A beam from an infrared laser mounted on a satellite sweeps the atmosphere below it, illuminating atmospheric particles moving with the wind. Because of their motion, the frequency of the infrared radiation scattered back to the satellite by the particles is shifted by an amount proportional to the wind speed (along the beam).

At present, measurements of wind speeds over the globe are sparse. Yet to know where the atmosphere is going one must know how fast and in which direction it is going there. Thus by making use of the Doppler effect, the global winds may some day be determined to an extent that will have a major impact on weather forecasting.

*✳15

All That's Best of Dark and Bright

There is measure in all things.
HORACE

A line from a poem by Byron gives the title to this chapter, although to say it encompasses *all* that's best of dark and bright would be a poetic exaggeration. A less poetic view of my subject was given by R. C. Hilborn in the *American Journal of Physics* (Vol. 52, 1984, p. 668) in which he opined that radiometry—the measurement of radiation—is an acronym for *r*evulsive, *a*rchaic, *d*iabolical, *i*nvidious, *o*dious, *m*ystifying, *e*xotic *t*erminology *r*egenerating *y*awns. This is understandable given the bewildering array of terms and units in radiometry and photometry. Photometric units in particular—*phot, stilb, nit, troland, nox,* and *skot*—are like characters in a fairy tale. The units *goblin* and *gremlin* would not be out of place in this list. It sometimes seems that the aim of radiometry and photometry is to coin as many terms and confect as many units as possible.

I shall say as little as possible about this welter of terms and units, but I must discuss a few of them to avoid misunderstanding. One misunderstanding seems widespread, which you can verify for yourselves.

Objects Near and Far

Ask your friends, without defining *brightness*, which everyone grasps intuitively if not securely (I used *brightness* without definition in previous chapters, yet I doubt that this caused any consternation), how the brightness of an object depends on its distance from an observer. When I pose this question to students, almost all of them say that brightness depends on the inverse square of the distance. In so doing, they are guided by a psychological rather than a physical principle: the desire to be as complicated as possible and thereby sound scientific. This desire can override common sense and commonsense perceptions. For if attenuation is negligible and the angular size of an object is not so small that the wave nature of light manifests itself explicitly (conditions satisfied more often than not by the objects of everyday life), its brightness is independent of distance. Examples of this abound, such as that shown in Figure 15.1, photographs of a fire taken at two different distances but with identical camera settings. The far fire (right) is as bright as the near one (left). Another example is the row of ceiling lights in Figure 4.4.

Aside from the desire to appear scientific, confusion about the dependence of brightness on distance arises because we are equipped with two detectors: our bodies taken as a whole are *irradiance* detectors, and embedded in them are two *radiance* detectors—our eyes. To understand these two radiometric terms we must grasp the concept of solid angle. This will be easier if we begin with planar angles.

Figure 15.1
As one moves away from a fire, it becomes less hot, but its brightness remains the same. These photographs were taken at different distances but with identical camera settings.

The Measure of Directions in the Plane and in Space

Planar angles are ratios of lengths. Imagine a directed line segment **OB** (see Figure 15.2) to be rotated about O into another direction **OC**. The angle between the two directions **OB** and **OC**, expressed in *radians*, is defined as the ratio of the length of the circular arc traced out by the tip of **OB** to the length of **OB**.

According to this definition, the angle between two perpendicular directions is $\pi/2$ radians. This is because the circumference of a circle is 2π times its radius, and rotating **OB** through a right angle takes it one-quarter of the way around a complete circle. But why this nattering about radians? What happened to good old-fashioned degrees?

We owe to Babylonian astronomers of over two thousand years ago our convention of dividing the circle into 360 parts, and I do not advocate that we should abandon it. I would dread

Figure 15.2
The angle between two directions **OB** and **OC** in the plane is the arc length s divided by the length of **OB**. Angle is a measure of a set of directions: the top right set is larger than the top left. In three dimensions, solid angle is the measure of a set of directions. The solid angle of a set of directions from O is the area A intersected by these directions on a sphere centered at O divided by the square of the distance from O to A. The bottom right set is larger than the bottom left set.

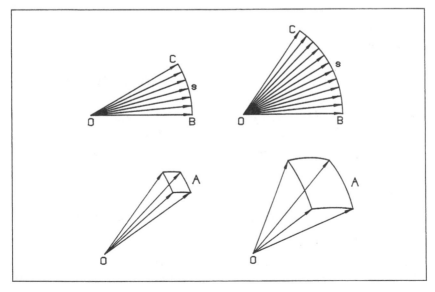

having to say in everyday conversation that an angle is 0.611 radian instead of 35 degrees. But when it comes to computations or understanding what an angle is, radians, not degrees, are my choice.

We may also interpret the angle between two directions **OB** and **OC** as the size (or *measure*) of the set of all directions beginning at *O* and ending on each point of the arc between *B* and *C*. This gives precision to notions about one set of directions being larger than another. Just as length is the measure of a set of points on a line and area is the measure of a set of points in the plane, angle is the measure of a set of directions in the plane. These measures allow us to arrange geometrical entities in order of increasing size; they arose not out of abstract mathematics but out of mundane affairs. Without the concept of area we could not distribute cropland equitably or determine its productivity.

Directions need not lie in a plane. We can easily imagine a bundle of directions originating from a point *O* and ending on the points of a surface instead of an arc. What is the measure of this bundle?

Proceeding by analogy with planar angles, we inscribe a sphere around the point *O*. A bundle of directions emanating from *O* intersects this sphere in an area (see Figure 15.2). The ratio of this area to the square of the radius of the sphere is defined as the solid angle of the bundle and is expressed in *steradians*, stereo being a combining form derived from the Greek *stereos* meaning solid or stiff (whence comes stereophonic, stereoscopic, etc.). The measure of the set of all directions in space is 4π steradians.

By the solid angle subtended by an object at a point is meant the solid angle of the bundle of directions emanating from it and ending on the object. For example, the solid angle subtended by the sun at any point on earth is about 0.00006 steradian.

From this mathematical frying pan, we are now better equipped to hop back into the fire.

Irradiance, Radiance, Luminance, and Brightness

Imagine a small, flat-plate radiation detector oriented in a fixed direction at some point in space. The rate at which radiant energy impinges on this detector from all directions in a hemisphere, divided by the detector area, is called the *irradiance* at the point; its metric units are watts per square meter. This radiant energy may be confined to a very narrow range of wavelengths (the irradiance is then qualified as monochromatic) or may include the electromagnetic spectrum from ultraviolet to infrared (e.g., the solar irradiance discussed in Chapter 10).

If the detector measures radiation from a source that is small compared with its distance, and if attenuation by the intervening medium is negligible, the irradiance is inversely proportional to the square of this distance. This is because the radiant energy crossing a spherical surface centered on the source is constant, but the area of the sphere increases as its radius squared. Hence the amount of radiant energy allocated to a unit area of the sphere must decrease inversely as the radius squared (see Chapter 10 for a more detailed explanation).

We are warmed by both visible and invisible radiation from a fire. As we move away from it, we are warmed less: the irradiance of the fire decreases and our detector (skin) has a fixed area. Yet, although our skin senses the fire as less hot, our eyes (or a camera) sense it as just as bright (see Figure 15.1). The radiant energy from the fire collected by a lens, be it in the human eye or in a camera, is inversely proportional to the distance squared, but so is the size of the image, hence the amount of light on that part of the retina (or film) illuminated by the flames remains the same. This is a specific example of the consequences of a general property of *radiance*: it does not vary with distance.

To define radiance, we imagine, as before, a small detector positioned at a point in space. But now we attach a collimator to the detector so that it receives energy from only a small set of directions close to the perpendicular to its surface. The radiance is defined as the rate at which the detector receives radiant energy divided by its area and the solid angle of the directions from which it receives this energy; the units of radiance are therefore watts per square meter per steradian. If the detector is oriented in a different direction, the radiance it measures generally changes (one exception being a whiteout, in which the radiance is *isotropic*: the same in all directions). Radiance depends on direction and completely characterizes the radiation field at a point. From radiances we can obtain irradiances but we cannot, in general, do the converse.

Radiance, unlike irradiance, does not change with distance (with suitable caveats). Although this was exemplified by the fire photographs, the invariance of radiance does not require the intervention of lenses and film. By the invariance of radiance with distance, I mean that the radiance in a particular direction does not change as we move along it.

To understand irradiance and radiance, it is best not to dwell too much on their definitions. An understanding of scientific terms is acquired more from seeing how they are used than from pondering their formal definitions, just as an understanding of words is acquired more from seeing how they are used than from studying dictionaries.

Snow illuminated by sunlight exemplifies the distinction between radiance and irradiance. We may pretend that snow perfectly reflects visible light, a realistic approximation for clean, fine-grained snow. Although the irradiances of the light from such snow and from the sun are the same, the radiances are vastly different. Direct sunlight is confined to a solid angle of about 0.00006 steradian, whereas sunlight reflected by snow is more or less uniformly spread into a hemisphere (about 6 steradians). Thus the radiance of direct sunlight is about 100,000 times that of snow.

I stated previously that the eye is a radiance detector. This was a lie told for expediency. Now I have to set the record straight: the eye is actually a *luminance* detector.

Light of each visible wavelength contributes differently to the sensation of brightness. This is shown in Figure 15.3, which depicts the *luminous efficiency* of the human eye. Two monochromatic sources of light, well separated in wavelength (e.g., red and blue), will appear different by virtue of their different colors regardless of their radiances. But suppose we compare two monochromatic sources with equal radiances and only slightly different wavelengths. For example, take them to be 650 and 670 nm, which correspond to nearly indistinguishable reds. According to the luminous efficiency curve (Figure 15.3), the 650 nm source will be brighter by about a factor of three than the 670 nm source, even though their radiances are equal.

Figure 15.3

The eye is not equally efficient at all wavelengths at converting radiant energy into the sensation of brightness.

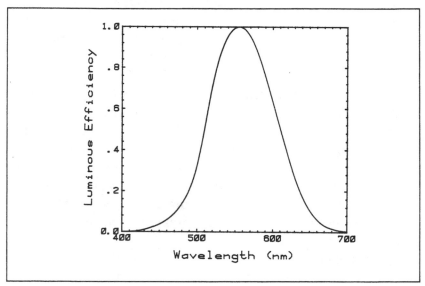

From the radiance spectrum of a source (the sun, the sky, a field of wheat) we can obtain the associated luminance: it is the sum of the radiances in each small wavelength interval weighted by the corresponding luminous efficiency (times a factor to convert radiance units to luminance units—about which the less said, the better). As a rough approximation, the luminance of a source often can be taken as proportional to its radiance at the wavelength of peak luminous efficiency (about 555 nm). Radiance is a radiometric quantity, luminance its photometric counterpart.

Now I must confess to another lie: our eyes do not detect luminance, but rather *brightness*. Radiance can be measured and is expressed in familiar units. Luminance is a more nebulous quantity, although at least it can be obtained from radiance by well-defined operations. But brightness does not have a nice operational definition, nor does it even have units. According to a recommendation made by the 1933 Committee on Colorimetry of the Optical Society of America, brightness is the "attribute of sensation by which an observer is aware of differences of luminance." This is how I use brightness, but before discussing the distinction between it and luminance, I have a few loose ends to tie.

Intensity is a term in widespread use, especially in textbooks. I used it in Chapters 2 and 3 (without defining it) but have avoided it in this one. Although intensity has a well-defined meaning, more often than not it is used according to a rule spelled out in *Alice in Wonderland*: "When *I* use a word," Humpty Dumpty said in rather a scornful tone, "it means just what I choose it to mean—neither more nor less."

Intensity has been, still is, and no doubt will continue to be used—including by me—to mean either irradiance, radiance, luminance, or even brightness, with its precise meaning determined by context. Sometimes it isn't worth the trouble to make fine distinctions. But we should recognize when they are called for. To say that the intensity of light from snow is the same as that from the sun is either approximately correct or wildly incorrect depending on whether by intensity is meant irradiance or radiance.

Everyone does not use radiometric and photometric terms in the same way. Astronomers seem to use *brightness* for radiance. Moreover, they study objects the radiances of which do depend on the inverse square of distance. Stars (with the exception of our sun) are effectively point sources. And the radiometric rules for point sources are different from those for extended sources, the kinds of objects that surround us every moment of our waking lives.

The Subjectivity of Brightness

To understand the world we see requires blending psychology with physics. Equipped only with a knowledge of optics, we would make predictions about what humans perceive that would be inaccurate or even ludicrous.

Brightness is sometimes essentially identical with luminance, but not always. This is shown in Figure 15.4. What is special about this drab winter scene? Any aesthetic value it may have is marred by the ugly wires stretching across it. Don't I know enough about photography to exclude wires? Yes, but in this instance, the wires are what is interesting. Notice how their brightness changes abruptly depending on backdrop. With the bright sky behind them, the wires are dark. But as their backdrop changes from bright sky to dark woods, the wires abruptly be-

Figure 15.4
Seen against the bright, overcast sky, the wires appear dark. But as their backdrop changes from sky to dark woods, the brightness of the wires abruptly increases. The luminance of the wires is approximately constant along their length, but their brightness changes discontinuously.

come brighter. This photograph was taken on an overcast day, hence there is no reason to expect the luminance of the wires, illuminated by more or less diffuse skylight, to change suddenly. Although the luminance of the wires may be constant along their length, their brightness is not. In general, we do not sense absolute luminances but rather differences of luminance. As another example, I often have seen falling snowflakes abruptly transformed from black to white as their backdrop changed from bright sky to dark woods.

Lest I mislead you, I must note that sometimes one or more overhead wires in a group are objectively brighter than the others, and sometimes the objective brightness of a wire does change over its length. Brightness differences may be objective—they arise from real luminance differences—or they may be creations of the eye-brain combination.

Perhaps more striking examples of the subjectivity of brightness are winter scenes such as that shown in Figure 15.5. The brightest object on this overcast day appears to have been the snow-covered roof. Under conditions of equal illumination, snow

Figure 15.5
The snow-covered roof in this scene appeared to be brighter than the overcast sky illuminating it. Yet snow cannot be brighter than the brightest part of the clouds that illuminate it.

on the ground is usually brighter than clouds, and this is how we perceive it even when it is not. On a sunny day, both clouds and snow are illuminated by the same direct sunlight, but when the sky is covered with thick clouds, snow is illuminated by cloud light. How can snow be brighter than its source of illumination? Objectively, it cannot; but subjectively, it often is.

We compare snow on the ground with surrounding darker objects such as trees or the horizon sky, which often is not as bright as the zenith sky on an overcast day. To observe snow on the ground and the zenith sky simultaneously requires the help of a mirror. On many a cloudy winter day when I would have sworn that my snow-covered lawn was brighter than the overhead sky, a mirror set in the snow convinced me otherwise (see Figure 15.6) even though mirrors are not perfectly reflecting. Try it yourself. You may be astounded by what you see.

Even with the mirror photograph in mind, you might find it difficult to accept that snow can never be brighter than the bright-

Figure 15.6
Mirrors can be used to show that the zenith sky on an overcast day is brighter than snow, despite appearances to the contrary, even though mirrors are not perfectly reflecting. The darker of the two mirror images is that of the horizon sky, part of the backdrop against which we often see snow.

est part of the overcast sky illuminating it. Surely, the snow grains can act as little lenses, thereby increasing the brightness. Although snow grains in a sense act as little lenses (not very good ones), they cannot increase the brightness of cloud light. Nor can *any* lens.

This statement may be greeted with howls from those who have used a magnifying glass to direct the sun's rays onto a sheet of paper or some hapless insect. I do not dispute that one can use a lens to burn paper by increasing irradiance. Nor do I dispute that focusing the sun's rays onto paper yields a spot much brighter than its surroundings. Indeed, the spot is brighter by approximately the ratio of the solid angle of the lens subtended at the spot to that of the sun. But neither the radiance of the spot nor that of the light illuminating it can be greater than that of the sun. This is a general result: lenses can increase irradiance but not radiance.

One way to convince yourself of this is to take an ant's point of view. An ant placed at the focal point of a lens in bright sunlight would quickly feel uncomfortable because of the increased irradiance: radiant energy is concentrated by the lens onto a smaller area. Looking upward through the lens, however, the ant would see a sun larger but no brighter than in the absence of the lens. The sun appears larger to him because he receives light from a larger set of directions. But radiance is energy divided by solid angle, so the increase in irradiance is compensated for by an increase in solid angle, and the radiance remains the same. Thus for the ant, the sun is hotter but just as bright, the lens serving to move the sun optically closer to him. This is similar to what happens when we move physically closer to a fire: it is hotter but just as bright.

I don't advise you to literally take the ant's point of view and look at the sun with your eye at the focal point of a magnifying lens. But you can do the following safer experiment. Cut two identical figures—discs, for example—from the same piece of white paper and place them on a dark background. Observe both figures simultaneously, but one through a magnifying lens. Because of reflection by the lens (a few percent), the magnified figure is likely to appear slightly less bright than its unmagnified twin. But depending on the (angular) size of these figures, the magnified one may appear slightly brighter.

To shed some light on this, I suggest a similar experiment. Again, cut two figures of the same shape but very different sizes (I have used rectangles because they are easy to form using scissors) from the same piece of white paper, place them on the same background, and illuminate them identically. Then vary your distance to the two figures and judge their relative brightness. I have tried this experiment with my wife, who was not biased by my expectations. At some distances, she perceived the smaller of the two rectangles to be less bright.

When I began this chapter with the assertion that the brightness of an object—a fire, for example—is independent of distance, I counted on your having an intuitive idea of what is meant by brightness. Now that we have explored this term in more detail, I must confess that although radiance and luminance are invariant with distance, brightness strictly is not. Previously, I stated that the invariance of radiance with distance (neglecting attenuation) is valid only when the wave nature of light does not manifest itself. Statements about the invariance of brightness with distance must be accompanied by an analogous physiological caveat. The brightness of an object does depend on the size of its retinal image, especially when the angular size of the object is small. Brightness also depends on pupil size and the state of adaption of the eye. We are all aware of the consequences of adaptation. When we turn out the lights in our bedrooms at night, we are plunged into darkness. But if we are still awake half an hour later, we notice how much brighter the room has become. Yet the luminance has remained the same. This is just one example among many of how all the tidy optical rules and laws are thrown into a cocked hat when we ask what human beings perceive rather than what instruments measure.

Selected Bibliography and Suggestions for Further Reading

Some references were woven into the text of this book, but sparingly to avoid giving it the musty odor of scholarship. The following list of books and papers serves to acknowledge further my indebtedness to predecessors and to help satisfy an appetite I hope to have whetted in my readers. I also include a few qualifications I chose not to make in the text.

General

Not surprisingly, the book that most closely resembles this one is my *Clouds in a Glass of Beer* (Wiley, 1987). My books are collections of essays in which I discuss in depth readily observable physical phenomena. Jearl Walker's unique *Flying Circus of Physics* (Wiley, 1977) is a collection of questions about many such phenomena, brief answers to them, and a long list of references to which readers are directed for more details. My books therefore complement his.

The only other popular science book I know about with windows as its theme is Elizabeth Wood's superb *Science from Your Airplane Window*, which was reprinted by Dover in 1975 but is unavailable.

Chapter 1

In discussing dew points, one must keep in mind that "the concept of a sharply defined temperature, at which condensation begins abruptly, represents only an approximation to the truth." This was pointed out by R. G. Wylie, D. K. Davies, and W. A. Caw in their paper "The Basic Process of the Dew-Point Hygrometer" in *Humidity and Moisture*, Vol. I, p. 125, Robert E. Ruskin, editor (Reinhold, 1965). These authors also note that "in dew-point hygrometry, the usual simple concept of condensation is inadequate when the accuracy sought is higher than about 0.2°C."

The inspiring story of Wilson Bentley, a farmer with no more education than that obtainable in rural Vermont of a century ago who nevertheless made original contributions to atmospheric science, has been told admirably by Duncan Blanchard ("Wilson Bentley: The Snowflake Man," *Weatherwise*, Vol. 23, 1970, p. 260; "Bentley and Lenard: Pioneers in Cloud Physics," *American Scientist*, Vol. 60, 1972, p. 746).

The details of frost formation, like those of ice crystal formation in the atmosphere, depend on temperature and vapor density, as evidenced by the diagrams in U. Nakaya's classic book *Snow Crystals, Natural and Artificial* (Harvard Univ. Press, 1954), in which he shows the different kinds of (artificial) ice crystals— dendrites, plates, needles, and so forth—formed in different environments. Observations of natural ice crystals are consistent with what Nakaya observed in the laboratory.

Frost formation on glass (or any other surface) is essentially a two-dimensional process, whereas ice crystal formation in the atmosphere is three-dimensional. Nevertheless, there are similarities between the two processes. Dennis Lamb, our ice crystal expert at Penn State, tells me that when environmental conditions favor the formation of dendritic snow crystals, he finds dendritic frost patterns.

Chapter 2

Interference colors in soap bubbles are discussed in *Soap Bubbles* by C. V. Boys, one of the classics of popular science and still available as a Dover reprint. These colors are also treated in *Soap Films* (G. Bell, 1929) by A. S. C. Lawrence.

Diffusion rings are discussed in detail by A. J. de Witte ("Interference in Scattered Light," *American Journal of Physics*, Vol. 35, 1967, p. 301).

Robert W. Wood discusses diffusion rings in his classic book *Physical Optics* (Macmillan, 1929)—the second edition (pp. 242–244), not the third, which was reprinted by Dover but again is out of print.

Chapter 3

Translations of the original paper setting down the Arago-Fresnel laws are found in W. F. Magie's compilation *A Source Book in Physics* (McGraw-Hill, 1935) and in *The Wave Theory of Light: Memoirs by Huygens, Young and Fresnel* edited by Henry Crew (American Book Company, 1900). Magie's book is the first place to look for hard-to-find scientific papers by our illustrious predecessors.

The advanced book from which I have learned the most about polarized light is William A. Shurcliff's *Polarized Light* (Harvard Univ. Press, 1962). At the popular level, I recommend Gunther P. Können's *Polarized Light in Nature* (Cambridge Univ. Press, 1985), which has many color plates.

The history of the development of sheet polarizing filters has been told engagingly by Edwin Land in "Some Aspects of the Development of Sheet Polarizers," *Journal of the Optical Society of America*, Vol. 41, 1951, p. 957.

Chapter 4

The ways in which the terms *Brewster angle* and *polarizing angle* have been used (and misused) are recounted by Akhlesh Lakhtakia in "Would Brewster Recognize Today's Brewster Angle?", *Optics News*, Vol. 15, June 1989, p. 14.

Chapter 5

Albert Rosenfeld's "Introducing Dr. Irving *Who*?" (*Saturday Review*, 4 March 1978, p. 46) is a short but good biographical sketch of Langmuir.

Harvey Leff, another connoisseur of light bulbs, has published an illuminating article in *The Physics Teacher* (Vol. 28, January 1990, p. 30): "Illuminating Physics with Light Bulbs."

Chapter 6

One of the best and most well-known popular articles on mirages is "Mirages" by Alistair B. Fraser and William H. Mach (*Scientific American*, January 1976, p. 102).

In a clever attempt to dispel the widespread misconception about the direct relation between refractive index and density, E. Scott Barr ("Concerning Index of Refraction and Density", *American Journal of Physics*, Vol. 23, 1955, p. 623) plotted the one versus the other for several organic compounds. He connected these data points so that the resulting curve answered his question in the figure caption, "Does index of refraction vary directly with density?": NO.

One chapter of Robert Greenler's superb book, *Rainbows, Halos, and Glories* (Cambridge Univ. Press, 1980), which is now available in paperback, is devoted to atmospheric refraction. Another good source on mirages (and much more besides) is M. Minnaert's *The Nature of Light and Color in the Open Air* (Dover, 1954).

For discussions of the moon illusion, see any good book on perception, such as Irving Rock's *Perception* (Freeman, 1984) or the book with the same title by Robert Sekuler and Randolph Blake (Knopf, 1985). An entire book devoted to this illusion, *The Moon Illusion*, edited by Maurice Hershenson, was published in 1989 by Lawrence Erlbaum Associates; the historical review by Cornelius Plug and Helen Ross and the annotated bibliography by Plug are by themselves worth the price of the book.

Everyone to whom I have shown David Jones's *The Inventions of Daedalus* (Freeman, 1982) has been enchanted by it. Jones has the remarkable ability to spin scientific tales that straddle the border between the plausible and the implausible. One of Daedalus's "plausible schemes" is the optically flat earth, one with a radius such that rays of light would "follow its curvature exactly," hence "people would not have realized that the Earth was round until they discovered that, with a good telescope, you could see the back of your head." The refutation of the notion that rays of light could travel around the world was given by two spoilsports, Craig F. Bohren and Alistair B. Fraser ("At What Altitude Does the Horizon Cease to be Visible?", *American Journal of Physics*, Vol. 54, 1986, p. 222).

Chapter 7

Despite the recent explosion of attention given to the greenhouse effect, its origins can be traced back almost two centuries. In

Worlds in the Making (Harper, 1908), Svante Arrhenius asserts "[t]hat the atmospheric envelopes limit the heat loss of the planets had been suggested about 1800 by the great French physicist Fourier."

In January 1990, "An Annotated Bibliography on Greenhouse Effect Change" by Mark David Handel and James S. Risbey was published by the Center for Global Change Science at the Massachusetts Institute of Technology. This bibliography is invaluable for anyone wishing to dig deeper into the greenhouse effect, especially its long history.

It hardly seems necessary to cite articles painting a gloomy picture of future catastrophies consequent upon carbon-dioxide-induced global warming. You will find plenty of such articles in newspapers and magazines chosen almost at random. For a dissident view, I recommend Richard S. Lindzen's "Some Coolness Concerning Global Warming," *Bulletin of the American Meteorological Society*, Vol. 71, 1990, p. 288. Perhaps this article marks the beginning of a long-overdue rational debate on the greenhouse effect. Lindzen is much too eminent in meteorology to be ignored.

In his article "Do Darker Objects Really Cool Faster?" *(American Journal of Physics*, Vol. 58, 1990, p. 244), Richard A. Bartels points out, as I did in my *Weatherwise* article on which Chapter 7 is based, that the cooling rates of black and white (as opposed to shiny) objects are essentially the same.

Chapter 8

The changing sounds of water as it is heated and comes to a boil are discussed on pages 88 and 89 of Sir William Henry Bragg's delightful little book *The World of Sound* (E. P. Dutton, 1942).

In this chapter I professed not to know the record for superheating water. This was said for literary effect; I must confess that it isn't true. I do know the record (279.5°C) because I read it in Robert C. Reid's article "Superheated Liquids" *(American Scientist*, March–April 1976, p. 146).

I first learned about the distinction between condensation temperature and boiling temperature in *Ebullimetric Measurements* by W. Swietislawski (Reinhold, 1945).

An article by Joseph Podzimek commemorating the hundredth anniversary of the publication of John Aitken's paper "On the Number of Dust Particles in the Atmosphere" was recently published in *Bulletin of the American Meteorological Society*, Vol. 70, 1989, p. 1538.

Chapter 9

The curves of temperature variations in time and space in this chapter were not drawn by whim. I computed these curves using an approximate method by T. R. Goodman: ("The Heat-Balance Integral and Its Application to Problems Involving a Change of Phase," *Transactions American Society of Mechanical Engineers*, Vol. 80, 1958, p. 335).

The use of low-emissivity coatings to prevent frost formation is discussed by I. Hamberg, J. S. E. M. Svensson, T. S. Eriksson, C. Granqvist, P. Arrenius, and F. Norin in "Radiative Cooling and Frost Formation on Surfaces with Different Thermal Emittance: Theoretical Analysis and Practical Experience," *Applied Optics*, Vol. 26, 1987, p. 2131. Note in particular Figure 9, which shows two cars that had been parked outdoors overnight, one with an ordinary windshield, the other with a windshield coated with a low-emissivity material.

Spectrally selective surfaces are discussed at an advanced level in Claes-Göran Granqvist's recent book *Spectrally Selective Surfaces for Heating and Cooling Applications* (SPIE Optical Engineering Press, 1989).

Chapter 10

A recent issue of *Journal of Geophysical Research* (Vol. 92, No. D1, 20 January 1987) contains a special section consisting of 12 papers on the variability of the solar constant.

For more on how solar radiation varies in space and time over the earth, I recommend *Physical Climatology* by William D. Sellers (University of Chicago, 1965).

Chapter 12

The thermal conductivity of gases is treated in any book on the kinetic theory of gases. At the popular level is H. G. Cowling's *Molecules in Motion* (Harper, 1960). Sir James Jean's *An Introduction to the Kinetic Theory of Gases* (Cambridge Univ. Press, 1940) is more advanced but does not make great mathematical demands on readers. An even more advanced text is Earle H. Kennard's *Kinetic Theory of Gases* (McGraw-Hill, 1938), from which I learned that the thermal conductivity of water vapor is slightly less than that of both nitrogen and oxygen (p. 180).

Chapter 14

A brief, but good, history of the life of Christian Doppler and the effect that bears his name, together with some of its modern applications, was given by Kurt Toman in "Christian Doppler and the Doppler Effect," *Eos, Transactions, American Geophysical Union*, Vol. 65, 1984, p. 1193.

An entire chapter of Hans Christian von Baeyer's *Rainbows, Snowflakes, and Quarks* (McGraw-Hill, 1984) is devoted to the Doppler effect, especially to its history and significance.

A short biography of Doppler is given in *Dictionary of Scientific Biography*, the 15 volumes of which are an excellent source for anyone seeking biographical information about scientists.

Chapter 15

I have learned much from Herschel W. Leibowitz's *Visual Perception* (Macmillan, 1965), a book filled with the insights of one of the leading authorities on perception. About half of this book is a collection of classic papers on visual perception.

Hermann von Helmholtz's *Popular Lectures on Scientific Subjects* are well worth reading, especially his lecture "The Eye as an Optical Instrument," in which he avers that "if an optician wanted to sell me an instrument which had all these defects [of the human eye], I should think myself quite justified in blaming his carelessness in the strongest terms, and giving him back his instrument."

Yves Le Grand's excellent book *Light, Colour and Vision* (2d ed., Chapman and Hall, 1968) is the first place I look for information on how the human eye responds to light.

Index